BURMA

LAOS

THAILAND

VIETNAM

CAMBODIA

SOUTH

CHINA

SEA

PENINSULAR
MALAYSIA

BRUNEI   SABAH

SARAWAK

SINGAPORE

INDIAN

OCEAN

INDO

20°

10°

0°

10°

90°   100°   110°

SOUTH-EAST ASIA

```
0        500      1000 KM
```

P   A   C   I   F   I   C

PHILIPPINES

O   C   E   A   N

A

S

E

NATURAL RESOURCES OF SOUTH-EAST-ASIA
General Editor: OOI JIN BEE

THE OFF-SHORE PETROLEUM RESOURCES OF SOUTH-EAST ASIA

THE OFF-SHORE PETROLEUM RESOURCES OF SOUTH-EAST ASIA
Potential Conflict Situations and Related Economic Considerations

CORAZÓN MORALES SIDDAYAO

Issued under the auspices of the
Institute of Southeast Asian Studies in Singapore

KUALA LUMPUR
OXFORD UNIVERSITY PRESS
OXFORD NEW YORK MELBOURNE

Oxford University Press

OXFORD LONDON GLASGOW
NEW YORK TORONTO MELBOURNE WELLINGTON
KUALA LUMPUR SINGAPORE HONG KONG TOKYO
DELHI BOMBAY CALCUTTA MADRAS KARACHI
NAIROBI DAR ES SALAAM CAPE TOWN

© *Oxford University Press 1978*
First published 1978
Second impression 1980

ISBN 978-0-1958-0488-1   ISBN 978-94-011-6855-7 (eBook)
DOI 10.1007/978-94-011-6855-7

*Published by Oxford University Press, 3, Jalan 13/3,
Petaling Jaya, Selangor, Malaysia*

A.M.D.G.

To
R.A.B.
and
my family

Acknowledgements

THIS study was conducted as part of the research programme of the Institute of Southeast Asian Studies on 'Oil Discovery and Technical Change in Southeast Asia'. It could not have been completed in its present form and at this time without the help of many people to whom I now want to express my deepest gratitude. It will not be possible to mention every one by name, but I feel I must give special mention to quite a few who have contributed their time or sources of information to help me improve my documentation as well as my understanding of the region, the petroleum industry, and the issues addressed in this study. It would be difficult to rank the degree of my indebtedness to each one, and the number of names involved requires a listing of those deserving special mention by geographical location or affiliation.

SOUTH-EAST ASIA

Brunei—I. Talog Davies, Attorney General.

Burma—U Nwai Tin, Coordinator, and Khin Han, Geophysicist, Martaban-Cities Service Inc.; and Esso Exploration Inc. staff, especially M.P. Gillert and B.H. Larramore.

Indonesia Dr. Hasyim Djalal, Ministerial Counsellor, Indonesian Embassy in Singapore; and S. Arief of Sritua Arief Associates, Jakarta.

Philippines—Dr. Emmanuel V. Tamesis, Chief Geologist, Philippine National Oil Company; Griselda Garcia-Bausa, Paleontologist, Petroleum Board; and Lido Gonzalo, School of Economics, University of the Philippines.

Thailand—Staff of the United Nations Development Programme, Technical Support for Regional Prospecting in East Asia (UNDP/CCOP), especially Dr. Frank F.H. Wang (USGS Geologist on special detail), Professor Laric V. Hawkins (Senior Marine Geophysicist), Dr. Thomas W.C. Hilde (then Senior Marine Geologist/Geophysicist), and Vinitha Thummanond (Administrative Officer); Jack A. Callow, Economic Affairs Officer, Energy Resources Section, United Nations ESCAP; staff members of the Thai Government's Department of Mineral Resources, especially Sangob Kaewbaidhoon and his staff; and Tammachart Sirivadhanakul, Chief, Regulatory Division of the National Energy Administration.

Singapore—The Asia Foundation; the American Resource Center; Charles Ahlgren, Commercial Attache, U.S. Embassy, and other officials; David Smallman, Second Secretary (Commercial), British High Commission; Allen G. Hatley, Manager, Cities Service East Asia; Esso Exploration Inc., especially George DeCoster, R.W. Murphy, and Maria Soh; *Petroleum News* managing editors (past and present), Jean MacDonald and Jim Matthews; Dr. John A. Diamond, Lecturer, Department of Economics, and Professor S. Jayakumar, Dean, Faculty of Law, University of Singapore; M. Rajaretnam, Research Officer, Institute of Southeast Asian Studies; and Dr. Max R. Langham, Research Officer, Agricultural Development Council.

UNITED STATES

Dr. Robert B. Helms, American Enterprise Institute, Washington, D.C.; Dr. Bension Varon and Dr. Valeriy Ovcharenko, United Nations, Centre for Natural Resources, Energy, and Transport; Julia F. Hutchins and Virginia Yates, Federal Energy Administration; William R. Hemphill, International Monetary Fund; John Foster and Myrna Celerian-Chunsanit, of the World Bank; Richard F. Meyer, Anny B. Coury, and Sherwood E. Frezon, of the U.S. Geological Survey in Reston and Denver; and petroleum staff of the U.S. Bureau of Mines.

Grateful acknowledgement is made to the following for permission to use illustrations and tables previously published: the National Petroleum Council, Washington, D.C., U.S.A., and *Petroleum News Southeast Asia*, Hong Kong.

Although already listed above, I would like to thank in an extra special way: Mr. Hatley for always being ready to answer any question or clarify obscure points; Dr. Wang for a genuine interest in providing data or documentation, and suggesting numerous other ways of improving the final product; Dr. Helms for helping me stay up-to-date with some of the theoretical and policy developments on the economic and other relevant issues in this study; Dr. Varon and Dr. Ovcharenko for their readiness to improve currentness of U.N. collected data and providing U.N. documents, and Mr. Foster, for keeping me in step with studies done at the World Bank.

The Asia Foundation provided financial support for travel and data collection. The trips through South-East Asia which resulted in some truly important contacts were made possible by this support.

I am also grateful in a special way to those who took the time to read and comment on the first draft of this study: namely, Professor Jayakumar, on the legal issues; Dr. Diamond and Dr. Langham, on the economic sections; Dr. Wang, Mr. Hatley, Salvador de Luna, and staff of the U.N. Centre on Natural Resources, Energy, and Transport, on the geological and other institutional aspects of the paper.

I also wish to express my thanks to the research staff of the Institute of Southeast Asian Studies for their encouraging comments. I especially appreciate the opportunity to test some of my ideas regarding presentation on Professor Eva M. Duka-Ventura of the University of the Philippines, a research fellow at the Institute while the study was developing in written form.

Last, but not least, I want to thank the support staff of the Institute of Southeast Asian Studies.

All shortcomings and errors, of course, remain the author's responsibility.

Institute of Southeast Asian Studies,   Corazón Morales Siddayao
Singapore,
16 March 1977

Contents

Tables

Figures

Abbreviations

API	American Petroleum Institute
ASCOPE	ASEAN Council on Petroleum
ASEAN	Association of South-East Asian Nations
BNFI	Bureau of Foreign Information (Manila)
CCOP	Committee for the Co-ordination of Joint Prospecting for Mineral Resources in Asian Offshore Areas
ECAFE	Economic Commission for Asia and the Far East
ESCAP	Economic and Social Commission for Asia and the Pacific
GDP	Gross Domestic Product
GDS	Geographically disadvantaged states
GNP	Gross National Product
ICJ	International Court of Justice
IMF	International Monetary Fund
LDC	Less developed countries
MB/D	Thousand barrels per day
NPC	National Petroleum Council (U.S.)
OGJ	*Oil and Gas Journal*
OPEC	Organization of the Petroleum Exporting Countries
PN	Petroleum News Southeast Asia
PRC	People's Republic of China
PREPF	Population, Resources, Environment and the Philippine Future
ROI	Return on investments
RP	Republic of the Philippines
SEAPEX	Southeast Asia Petroleum Exploration Society
SPE	Society of Petroleum Engineers
UN	United Nations
UNDP	United Nations Development Programme
USBM	United States Bureau of Mines
USGS	United States Geological Survey

Introduction

INTEREST in the off-shore petroleum[1] resources of South-East Asia was manifested in the 1960s when development in off-shore technology allowed oil companies to search beyond prospective land areas. The dramatic increases in oil prices in the early 1970s but more particularly the events of 1973 and 1974, when world oil prices were quadrupled by the oil exporting nations and major supply cutbacks were experienced by certain developed nations, further heightened this interest. Cost/price relationships had not only improved and made off-shore oil in hitherto less attractive areas commercially prospective; nations that were net importers and whose international exchange reserves were strained by the high import costs of foreign oil also found it prudent to begin looking for indigenous resources and to encourage such search.

The search for and discovery of petroleum in South-East Asia on the scale in which it has been conducted in the last ten years was new to the region. It was natural, therefore, for students of South-East Asia to raise questions about its progress, questions concerning international relations, social impacts, and economic policy implications. The purpose of this study is to try and answer the question: 'What are the potentials for conflicts or cooperation among nations arising from the search for petroleum resources in the seabeds of South-East Asia?'

The problem of conflicts or cooperation among nations is a topic that has many facets and may involve a multitude of issues, for example, legal, economic, technical, security, social, etc. In relation to off-shore petroleum resources, conflicts, although essentially legal in origin, have economic aspects. This study has, therefore, been oriented to the aspects more familiar to an economist, and has two major thrusts. First, the study discusses some relevant issues and identifies actual or potential conflict situations in the search for and development of South-East Asia's petroleum resources. Second, it seeks to identify the major relevant economic variables on which conflict situations are likely to have an impact, and the policy implications of the need to achieve maximum economic benefit from the presence of these resources.

FRAMEWORK OF ANALYSIS

The problem of conflict/cooperation arising from the development of South-East Asia's off-shore petroleum resources is studied within the framework of the demand for and supply of these resources. Potential conflict (or the need for cooperation) among nations arises only if (1) petroleum has value as a commodity and therefore a demand and (2) if petroleum may be found in South-East Asia's sedimentary strata, so that there is a potential indigenous supply in certain areas to which each nation can lay a claim. Given both, disputes would then essentially arise from conflicting claims to property rights over such resources, because of (1) incompatibility in perceptions of the equity of existing or proposed legal definitions of jurisdictional boundaries among nations concerned; (2) incompatibility of jurisdictional boundaries so delineated with geological and environmental phenomena; and (3) conflicting historical ownership claims.

Given the value of petroleum to the economic growth of each nation, access to such resources is manifestly desirable. International legal disputes over ownership of such resources constitute impediments to access, and therefore to development and utilization of those resources. Thus, maximization of economic welfare — the goal of every government — requires consideration of the costs and benefits of protracted disagreements. Petroleum available in twenty years has less social value than petroleum available in five years. As in the case of all scarce resources, trade-offs are faced — to give up a little of something in order to gain more of another.

DATA AND SCOPE OF STUDY

For purposes of this study, South-East Asia is defined in the conventional manner and includes the following countries: Brunei, Burma, Cambodia, Indonesia, Laos, Malaysia, the Philippines,

Singapore, Thailand and Vietnam.[2] The economic and resource data used are generally those published by national governments or submitted by such governments to the United Nations and similar international agencies. Where no government data are available, the usual industry sources are used, for example, crude oil production and other industry statistics in the *Oil and Gas Journal*, *Offshore* magazine, publications of the American Petroleum Institute and the U.S. National Petroleum Council, and releases by the Philippine Petroleum Association. Some of the historical data and projections have been estimated by the author.

Quantitative information will be expressed in metric units with some exceptions. Where the datum was originally reported in an imperial unit and where some loss of useful information might result from conversion, the original form will be retained and the metric equivalent will, in general, be included in parenthesis. (The exception to the latter will be 'barrels' which is still commonly used in the oil industry literature and which will therefore not be converted into metric tons where it is not necessary to do so for consistency in comparison.) Also, in the existing and proposed changes in the Law of the Sea, lengths are expressed in miles rather than in kilometres, while depths are expressed in metres. For simplicity, the use of this combined form will be retained in this study in discussing the boundary issues.

Quantitative analysis is confined to data readily available from published sources. In the absence of data on the region, the analysis of economic impacts and social costs and benefits has largely remained at theoretical levels and general discussions based on experiences in other regions. A major and longer study would be required to treat that subject in depth, and to arrive at empirical conclusions.

Chapter I discusses the relationship of energy to economic growth, the patterns of consumption of energy sources in South-East Asian countries, the import dependence of such countries, projections of future consumption, and the value of indigenous petroleum resources to these countries. Chapter II presents the potentials of petroleum production in South-East Asia. Chapters III, IV, and V analyse the three basic sources of conflict in South-East Asia that relate to petroleum resource development. Chapter VI analyses some economic and other impacts of petroleum resource development, and explores the economic welfare implications of disagreements versus cooperation. Chapter VII suggests certain policy implications and conclusions.

This study was completed in late 1976, and the statistical tables are based on data available in mid-1976. Minor revisions were made as the book went to press, mostly to accommodate later developments in the Law of the Sea. The detailed analyses, however, reflect the date of the study's completion.

1. In this book, the term 'petroleum' will be used to refer to both oil and/or gas inasmuch as exploration drilling for petroleum may result in discovery of either oil or gas or both.

2. Where necessary, the terms 'North Vietnam' and 'South Vietnam' will be used to refer specifically to matters related to or actions taken by the governments of these political divisions prior to reunification.

PART ONE

The Demand-Supply Situation

I

The Value of Petroleum to South-East Asian Economies

THE value of South-East Asia's petroleum resources lies basically in the value of energy to the region's economic activity and the changing pattern of energy consumption in the direction of increased dependence on petroleum. It is generally acknowledged that energy consumption is related to economic growth as measured in increases in the national output, although the precise relationship varies according to several factors.

The Relation between Energy and Economic Growth

Modern world history has shown that economic advancement, measured in terms of increased output, has been most rapid during the periods when energy-fed capital goods have been substituted for human-animal energy in production. The Industrial Revolution was the product of two great forces: development of the energy potential of coal and development of the machinery that used this energy. The demand for fuel and power has grown as a result of population growth, increasing general mechanization, a way of life based more and more on large and regular supplies of fuel and power to save time or increase comfort, and new uses of energy that have developed with the availability of relatively cheap and easily transportable oil supplies.

The rate of growth of energy consumption in developing countries has been projected to be higher than that of already developed economies. The potential for such growth is basically related to the potential for increasing the employment of electrical and mechanical energy in industrial and other economic applications. In fact, in the early stages of economic development, the energy/GNP ratio would tend to follow an upward trend.

While historical data show no consistent pattern in the relation-ship between GNP growth and energy demand growth, a strong relationship between the two does exist and the GNP growth rate must be given its proper weight in assessing the importance of energy to the economic development of any country or region. The precise relationship is dependent on many factors. These factors include the state of technological development of the economy, advances in thermal efficiency, structural changes in the economy that lead to the increasing importance of a sector that may either be energy-intensive (in which case it affects the ratio upward) or not energy-intensive (which pulls the ratio down), or changes in the overall pro-ductivity of labour and capital. The problem of precision in inter-pretation is especially acute for very short period changes.

ENERGY/GNP PER CAPITA RATIOS IN SOUTH-EAST ASIA

Table 1.1 shows a correlation of changes in consumption of energy per capita to changes in real GNP per capita for South-East Asian countries. Between 1965 and 1973 the ratio of changes in energy per capita and GNP per capita[1] were all greater than or equal to unity (1.0 to 7.57 for direct relationships).[2] The 1974/73 ratios, however, show both direct and indirect relationships; three inverse relationships ranged from −0.62 in Indonesia to −11.25 in Thailand.[3]

It will be noted that during the period 1965 to 1973 all but the Indo-chinese countries experienced per capita increases in energy consumption at rates higher than per capita GNP growth. In 1974, however, several countries reduced per capita consumption, suggest-ing an apparent shift in the relationship between energy consump-tion and GNP growth. Thus, while energy is necessary to output growth, some of its use may not necessarily be connected with pro-ductive activity. A valid partial explanation, of course, is that the negative effects on energy consumption of higher prices, especially of oil, were large in the non-producing sectors. Unfortunately, a breakdown of consumption by sectors is not available for most countries. Data from the Philippines show that all sectors reduced consumption of oil in 1974 but the significant decreases occurred in the transportation and power generation sectors.[4]

TABLE 1.1
SOUTH-EAST ASIA: GROWTH RATES OF ENERGY CONSUMPTION PER CAPITA AND REAL GNP PER CAPITA WITH RATIOS (COEFFICIENT FACTORS)

Country Variable	Average annual growth rates (%)		Country Variable	Average annual growth rates (%)	
	1973/65	1974/73		1973/65	1974/73
BRUNEI			PHILIPPINES		
Energy per capita	28.0	19.2	Energy per capita	4.5	0
GNP per capita	3.7	N.A.	GNP per capita	2.6	2.2
(Coefficient factor)[a]	(7.57)	N.A.	(Coefficient factor)	(1.73)	(--)
BURMA			SINGAPORE		
Energy per capita	2.6	-6.7	Energy per capita	10.1	3.2
GNP per capita	0.7	2.7	GNP per capita	9.4	5.3
(Coefficient factor)	(3.71)	(-2.5)	(Coefficient factor)	(1.07)	(0.6)
CAMBODIA			THAILAND		
Energy per capita	-5.2	-43.3	Energy per capita	11.7	-4.5
GNP per capita	-5.2	0	GNP per capita	4.5	0.4
(Coefficient factor)	(1.0)	--	(Coefficient factor)	(2.6)	(-11.25)
INDONESIA			VIETNAM, SOUTH		
Energy per capita	4.9	-3.7	Energy per capita	13.4	-48.4
GNP per capita	4.5	6.0	GNP per capita	-0.7	-0.4[b]
(Coefficient factor)	(1.09)	(-0.62)	(Coefficient factor)	(-19.14)	(121)
LAOS			VIETNAM, NORTH		
Energy per capita	11.4	-21.7	Energy per capita	-1.0	16.6
GNP per capita	2.5	5.9[b]	GNP per capita	-0.5	7.2[b]
(Coefficient factor)	(4.56)	(3.68)	(Coefficient factor)	(2.0)	(2.3)
MALAYSIA					
Energy per capita	6.2[c]	1.1[c]			
GNP per capita	3.7	4.5			
(Coefficient factor)	(1.68)	(0.24)			

Some Developed Economies:

UNITED STATES			JAPAN		
Energy per capita	3.3	-3.5	Energy per capita	9.1	-2.6
GNP per capita	2.5	-2.4	GNP per capita	9.6	-8.8
(Coefficient factor)	(1.32)	(1 46)	(Coefficient factor)	(0.95)	(0 3)

Sources: United Nations, *World Energy Supplies*, Stat. Series J, No. 19 (New York, 1976); World Bank, *Atlas: Population, Per Capita, and Growth Rates* (Washington, D.C. 1975); International Monetary Fund, *International Financial Statistics* (Washington, D.C., July 1976).

[a] Coefficient factor equals energy per capita growth rate divided by GNP per capita growth rate.

[b] Estimated: market GNP deflated by 1.102.

[c] Estimated: non-weighted average of three states.

Note: Growth rates for energy/capita computed at kg/capita; rates for GNP computed at US$ per capita.

LIMITATIONS/USEFULNESS OF ANALYSIS

The analysis is not without biases. First, energy consumption growth could be strongly influenced by activities not directly related to production. For example, increased single-passenger car use or increased military transport use could push up energy consumption change faster if these uses caused total transport use to grow at a rate faster than those of industrial or agricultural uses.[5] This may be true of the Philippines, Singapore, and Thailand. By using GNP per capita and energy per capita, the population bias on growth for the two types of data is at least reduced.[6]

Also, there are other variables that affect energy consumption, the most important of which is the price of the dominant fuel and the degree of dependence on this fuel. Sophisticated models of income elasticities may present better estimates, as they can take into account different explanatory variables. Still, the results are dependent on the time frame of the study, the level of aggregation of the data, the reliability of the data, and the variables included in the estimation. For example, the United Nations, using a simple model for the years 1950 to 1968, estimated the income elasticities of developing countries to be 1.6597 for all energy sources and 1.7517 for liquid fuels.[7] A later model of the World Bank, applied to non-OPEC developing countries for the years 1960 to 1973, yielded income elasticities for oil of 0.320 for low income countries, 0.403 for middle income countries, and 0.557 for the high income group.[8]

Also, as Samuelson points out, there is nothing sacred about elasticities. The bases are essentially arbitrary, and while elasticity expressions are invariant under changes of scale, they are not invariant under changes of origin.[9]

The statistical relationships are, nevertheless, useful for projecting the general direction of consumption growth for certain types of economies and the growth magnitudes that may be expected. Such information is necessary to policymakers and to policy analysts. For the purpose of this study it is useful to observe the relationship between energy growth and GNP growth because it provides a guide for energy resource policy.

Energy and Petroleum Consumption in South-East Asia

The patterns of consumption of different fuels are essentially determined by the cost of alternative fuels, the availability of such fuels, and developments in technology that determine the intensity with which a specific fuel will be used. The latter is, in turn, influenced by the long-run availability and cost of a fuel; in the case of petroleum its greater availability at costs relatively much lower than coal influenced the rapid development of oil-based technology and the shift towards increasing oil consumption, not only in Asia but world-wide. Over a period of time longer than the short-run, therefore, the factors which affect demand for an energy resource depend on the existence of readily available alternative sources of supply. In the South-East Asian countries, for practical purposes, no alternative to oil exists for any foreseeable period of time, and the demand for oil may be relatively inelastic with reference to price increases.[10]

As the growth of energy, and particularly of oil consumption world-wide and in South-East Asia, actually evolved from changes in the energy supply picture, it will be useful to summarize post-Second World War developments in the international oil market before discussing energy consumption patterns in the region.

THE POST SECOND WORLD WAR MARKET FOR OIL

From the late 1950s to about the summer of 1970, world oil sales were characterized by a continuous decline in both current and real oil prices from the peak reached during the closing of the Suez Canal in 1957. It was a period of abundant low-cost petroleum supplies, resulting from several factors that included the following: (1) rapid development of new oil resources in Venezuela, North Africa, and the Middle East as new oil exploration groups sought rapid paybacks on their investments in new discoveries; (2) import restrictions in the United States beginning in 1957 which had reduced export opportunities to that major buyer; (3) growth in world proved recoverable reserves of crude oil at rates faster than world consumption; and (4) low freight rates on international shipments.

TABLE 1.2

CONSUMPTION TRENDS OF COMMERCIAL ENERGY IN SOUTH-EAST ASIA:1963, 1973, 1974

(Volumes in thousand tons of coal equivalent, except where otherwise shown)

Country	Form of energy	1963		1973		1974		Average annual rate of growth (%)	
		Consumption[a]	Per cent of total[a]	Consumption[a]	Per cent of total[a]	Consumption[a]	Per cent of total[a]	1973/1963	1974/1973
WORLD	Coal	2 171 490	46.1	2 502 904	31.6	2 530 733	31.8	1.4	1.1
	Liquid fuels	1 680 873	35.7	3 597 884	45.5	3 566 777	44.8	7.9	-0.9
	Natural gas	757 023	16.1	1 622 726	20.5	1 668 207	20.9	7.9	2.8
	Hydro-electricity	100 192	2.1	185 775	2.4	205 098	2.6	6.4	10.4
	Total	4 709 577	100.0	7 909 289	100.0	7 970 815	100.0	5.3	0.8
	Per capita (kg.)	1 490		2 080		2 059		3.4	-1.0
FAR EAST,[1] DEVELOP-ING COUNTRIES	Coal	81 291	61.0	97 273	39.1	106 749	41.7	1.8	9.7
	Liquid fuels	43 860	32.9	126 428	50.9	121 911	47.6	11.2	-3.6
	Natural gas	5 662	4.2	18 621	7.5	20 658	8.1	12.7	10.9
	Hydro-electricity	2 630	2.0	6 302	2.5	6 543	2.6	9.1	3.8
	Total	133 143	100.0	248 624	100.0	255 861	100.0	6.4	2.9
	Per capita (kg.)	154		228		229		4.0	0.4
BRUNEI	Coal	—	—	—	—	—	—	—	—
	Liquid fuels	33	15.4	101	5.9	112	5.3	11.8	10.9
	Natural gas	181	84.6	1 610	94.1	1 998	94.7	24.5	24.1
	Hydro-electricity	—	—	—	—	—	—	—	—
	Total	214	100.0	1 712	100.0	2 110	100.0	23.0	23.2
	Per capita (kg.)	2 276		11 805		14 067		17.9	19 9

BURMA	Coal	217	18.2	116	6.5	67	4.0	-6.0	-42.2
	Liquid fuels	922	77.5	1 577	89.1	1 534	90.9	5.5	-2.7
	Natural gas	21	1.8	20	1.1	27	1.6	-0.5	35.0
	Hydro-electricity	29	2.4	59	3.3	57	3.4	7.3	-3.4
	Total	1 190	100.0	1 771	100.0	1 685	100.0	4.1	-4.9
	Per capita (kg.)	50		60		56		1.9	-6.7
CAMBODIA	Coal	NA[b]	NA	NA[b]	NA	NA[b]	NA	NA	NA
	Liquid fuels	292	NA	226	NA	131	NA	-2.5	-42.0
	Natural gas	—[c]	—	—[c]	—	—[c]	—	—	—
	Hydro-electricity	NA	NA	NA	NA	NA	NA	NA	NA
	Total	NA		NA		NA		NA	NA
	Per capita (kg.)	50		30		17		-5.0	-43.3
INDONESIA	Coal	601	5.5	157	0.8	161	0.8	-12.5	2.6
	Liquid fuels	6 457	59.4	13 595	65.9	12 341	60.7	7.7	-9.2
	Natural gas	3 727	34.3	6 701	32.5	7 635	37.6	6.1	13.9
	Hydro-electricity	86	0.8	172	0.8	191	0.9	7.2	11.1
	Total	10 871	100.0	20 625	100.0	20 327	100.0	6.6	-1.4
	Per capita (kg.)	108		164		158		4.2	-3.7
LAOS	Coal	—	—	—	—	—	—	—	—
	Liquid fuels	76	100.0	254	96.2	203	95.8	12.8	-20.1
	Natural gas	—	—	—	—	—	—	—	—
	Hydro-electricity	(1969 = 2)	—	9	3.4	10	4.7	46.0	11.1
	Total	76	100.0	264	100.0	212	100.0	13.4	-19.7
	Per capita (kg.)	30		83		65		10.7	-21.7

Country	Form of energy	1963		1973		1974		Average annual rate of growth (%)	
		Consumption[a]	Per cent of total[a]	Consumption[a]	Per cent of total[a]	Consumption[a]	Per cent of total[a]	1973/1963	1974/1973
MALAYSIA	Coal	34	1.3	48	0.8	51	0.8	3.5	6.2
	Liquid fuels	2 412	94.1	5 848	94.0	6 042	93.8	9.2	3.3
	Natural gas	73	2.8	186	3.0	200	3.1	9.8	7.5
	Hydro-electricity	43	1.7	138	2.2	148	2.3	12.4	7.2
	Total	2 562	100.0	6 220	100.0	6 441	100.0	9.3	3.6
	Per capita (kg.)	NA[d]		NA[d]		NA[d]			
PHILIP-PINES	Coal	171	2.9	48	0.4	62	0.5	-22.0	29.2
	Liquid fuels	5 551	94.0	11 928	96.0	12 199	95.4	8.0	2.3
	Natural gas	—	—	—	—	—	—	—	—
	Hydro-electricity	184	3.1	445	3.6	529	4.1	9.2	18.9
	Total	5 906	100.0	12 421	100.0	12 790	100.0	7.7	3.0
	Per capita (kg.)	197		309		309		4.6	0.0
SINGAPORE	Coal	8	0.7	5	0.1	6	0.1	-4.5	20.0
	Liquid fuels	1 096	99.3	4 356	99.9	4 565	99.9	14.8	4.8
	Natural gas	—	—	—	—	—	—	—	—
	Hydro-electricity	—[e]		—		—		—	—
	Total	1 104	100.0	4 362	100.0	4 572	100.0	14.7	4.8
	Per capita (kg.)	620		1 996		2 060		12.4	3.2
THAILAND	Coal	47	1.9	135	1.1	194	1.6	11.1	43.7
	Liquid fuels	2 404	98.0	12 116	96.9	11 809	95.8	17.6	-2.5
	Natural gas	—	—	—	—	—	—	—	—
	Hydro-electricity	—		252	2.0	322	2.6	24.5	27.8
	Total	2 452	100.0	12 503	100.0	12 325	100.0	17.7	-1.4
	Per capita (kg.)	85		314		300		14.0	-4.5

VIETNAM. SOUTH	Coal	113	11.2	16	0.2	10	0.2	-17.5	-37.5
	Liquid fuels	894	88.5	8 096	99.8	4 247	99.7	24.5	-47.5
	Natural gas	—	—	—	—	—	—	—	—
	Hydro-electricity	2	0.2	2	0.03	2	0.05	0.0	0
	Total	1 010	100.0	8 114	100.0	4 259	100.0	23.0	-47.5
	Per capita (kg.)	66		407		210		20.0	-48.4
VIETNAM. NORTH	Coal	2 409	91.4	2 785	81.3	3 330	81.6	1.5	19.6
	Liquid fuels	200	7.6	593	17.3	702	17.2	11.5	18.4
	Natural gas	—	—	—	—	—	—	—	—
	Hydro-electricity	26	1.0	49	1.4	49	1.2	6.5	0.0
	Total	2 635	100.0	3 427	100.0	4 082	100.0	2.7	19.1
	Per capita (kg.)	147		151		176		0.3	16.6

	1963	1973	1974
W. Malaysia	288	552	556
Sabah	203	545	580
Sarawak	339	530	508

Source: United Nations, *World Energy Supplies, 1950-1974.* Stat. Series J. No. 19.

a Detail may not add up to total because of rounding.

b Data showed consumption of 12 million metric tons of coal in 1964, and 5 million metric tons in 1972.

c Some hydroelectric power was consumed between 1967 and 1971.

d A per capita computation for the Federation of Malaysian States is not available, as data presented are given for W. Malaysia, Sabah and Sarawak separately. The breakdown is as follows:

e Net trade of hydroelectric power is shown as negative in the report, but this appears to be in error from verification locally.

1 Includes besides the 10 South-East Asian countries, Bangladesh, Sri Lanka, India, and Korea.

Notes: — = zero or not applicable   NA = not available

The buyer's market came to a dramatic end in mid-1970, and the declining price trend reversed. As world dependence on oil accelerated, while discovery of new major oil reserves lagged behind consumption rates, the demand/supply balance tilted in favour of the producing nations. In 1970 tax rates on production were raised by the Middle East and North African states. In the next three years, several price increase agreements were signed between the oil companies and the host nations that reflected the monopolistic position of the latter.

In October 1973 unilateral increases that quadrupled pre-1973 prices, along with cutbacks in production in most Middle East countries and supply embargoes on certain consuming countries, ended world complacency with regard to oil consumption and production. Most severely affected were, of course, non-producing developing countries which were almost totally dependent on petroleum for their economic development programmes. These countries, at the same time, had limited foreign exchange resources to cope with the sudden sharp increase in the cost of oil imports.

SOUTH-EAST ASIA'S DEPENDENCE ON OIL

The rate of growth of oil consumption in Asia has been higher than the rest of the developing world. During the thirty years preceding the 1960s, growth in the Far East's oil consumption averaged at about 5 per cent per year.[11] Between 1960 and 1970 this was estimated at 9.4 per cent per year.[12] In specific South-East Asian countries the average annual increase during the same period ranged from 8.9 to 15.6 per cent.[13] Only part of the increase was attributed to economic growth and part of the demand growth was a result of substitution of oil for coal.[14]

Table 1.2 shows consumption trends of all forms of commercial energy in South-East Asian countries from 1963 to 1974 based on U.N. data. Growth rates were computed for the 10-year period 1963–73; changes in the year 1974 relative to 1973 are also shown.

Before discussing the trends shown in the table, it will be appropriate to point out that the data have two main shortcomings: (1) The data cover only 'commercial' energy, i.e., fuels that are traded and for which records are available, allowing consumption estimates.

Fuels such as wood, bagasse, paddy husks, and sawdust,. which in some countries constitute a significant source of energy,[15] are therefore excluded. (2) Countries with oil producing capabilities, both in the upstream or downstream stages, tend to introduce biases in their consumption levels. In the production of crude oil in countries such as Indonesia and Brunei, equipment are petroleum-fed and field use of oil accounts for a significant percentage of fuel used in industry. In a country such as Singapore, with a refining sector contributing over one-third of the country's total industrial output, a similar bias could arise because of the high energy/output ratio inherent in the refining process.[16] Nevertheless, the data indicate orders of magnitude that are useful for comparison and are meaningful in terms of the changing importance of individual fuel groups.

Table 1.2 shows how the importance of coal as a source of energy dropped between the years 1963 and 1973. This was true worldwide, in Far East LDCs,[17] and in all South-East Asian countries where coal was used. While the contribution of coal increased in 1974 in Far East LDCs as that of petroleum dropped in absolute and relative terms, the share of coal in total energy consumption in South-East Asian countries continued to decline. In eight South-East Asian countries, excluding Brunei and Indonesia, oil has supplied 95 per cent of energy requirements. In Brunei and Indonesia, long-time petroleum producers, oil and natural gas met close to 100 per cent of energy requirements during the years 1963, 1973, and 1974. Between 1963 and 1973, the average annual growth rate of oil consumption in South-East Asian countries ranged from 5.5 per cent (Burma) to 24.5 (Brunei). (Consumption in Cambodia declined at a − 2.5 per cent rate.)

'The dramatic increase in world oil prices in 1973 generally reduced the rate of consumption in 1974. Except for Brunei and North Vietnam, which showed increases of 24 and 18 per cent respectively, all other South-East Asian countries showed either negative growth or slower increases. Even Indonesia, the major producer, showed a drop in consumption of − 9.2 per cent. At the same time, Indonesia, the Philippines, and Singapore increased their consumption of coal, the first two doing so with indigenous resources.[18]

The growth of oil consumption in both developed and developing countries in the next ten years is generally expected to be at rates lower than those experienced during the period of relatively cheap oil and natural gas supplies. As incomes and population grow, demand for energy will also continue to grow, but in 1976 the World Bank projected energy growth in LDCs at only 5 per cent.[19] With the sharp increases in oil prices, countries which have relied on imported oil have found a need to adjust their energy use per unit of output. These price increases have also intensified exploration for new indigenous supply sources.

IMPORT DEPENDENCE OF SOUTH-EAST ASIAN COUNTRIES
FOR PETROLEUM REQUIREMENTS

Table 1.3 shows data on consumption of petroleum products against indigenous crude oil production in South-East Asian countries, for the years 1963, 1973, and 1974. It also shows the resulting import requirements. The three major oil producing countries, which are also net exporters, are grouped under one heading, and the net importing countries under another heading.[20]

Oil moves internationally in and out of the region. For example, Malaysia still consumes some oil from the Middle East while it exports a substantial part of its low-sulphur crude production.[21] One reason is that 'sweet' crude commands higher prices overseas than what Malaysia would pay for imported crude with higher but environmentally acceptable sulphur content. Another reason is that market demand and local refinery design cannot accommodate all types of domestically produced crude. For example, oil from Sarawak is light crude, so that very little heavy residual fuel is processed from it.[22]

Overall, the picture is one of a surplus for the region. In reality, however, this does not help the individual country that depends on imported oil. Only two of the seven net importers have crude oil production, and only one of these two is almost self-sufficient. Burma's production supplied as much as 90 per cent of the country's requirements in 1974, but the eight other countries are totally dependent on imports for their petroleum requirements. (Thailand's production is negligible.) Reports have indicated that Burma was self-sufficient in 1975,[23] and that it would be a net exporter by the fiscal year 1976-7.[24]

TABLE 1.3
CONSUMPTION OF PETROLEUM PRODUCTS AND IMPORT DEPENDENCE, 1963-74 (SELECTED YEARS)
(Volumes in thousand metric tons)

Countries	1963 Consumption (1)	1963 Crude production (2)	1963 Estimated net imports (exports) (3)	1963 Net imports as % of cons. (4)	1973 Consumption (5)	1973 Crude production (6)	1973 Estimated net imports (exports) (7)	1973 Net imports as % of cons. (8)	1974 Consumption (9)	1974 Crude production (10)	1974 Estimated net imports (exports) (11)	1974 Net imports as % of cons. (12)
NET EXPORTERS												
Brunei	21	3 438	(3 417)	—	64	11 053	(10 989)	—	71	928	(857)	—
Indonesia	4 150	22 381	(18 231)	—	8 691	66 216	(57 525)	—	7 893	64 460	(56 567)	—
Malaysia	1 580	52	1 528	96.7	3 840	4 340	(500)	—	3 968	3 844	124	3.1
W. Malaysia	1 387	0	—	—	3 332	—	—	—	3 435	—	—	—
Sabah	65	0	—	—	270	—	—	—	303	—	—	—
Sarawak	128	52	—	—	238	4 340	—	—	230	3 844	—	—
NET IMPORTERS												
Burma	595	656	(41)	—	1 024	968	56	5.5	992	888	104	10.5
Cambodia	188	—	188	100.0	147	—	147	100.0	85	—	85	100.0
Laos	48	—	48	100.0	162	—	162	100.0	129	—	129	100.0
Philippines	3 606	—	3 606	100.0	7 757	—	7 757	100.0	7 933	—	7 933	100.0
Singapore	719	—	719	100.0	2 880	—	2 880	100.0	3 091	—	3 091	100.0
Thailand	1 565	6	1 559	99.6	7 927	39	7 888	99.5	7 757	10	7 727	99.9
Vietnam, South	581	—	581	100.0	5 214	—	5 214	100.0	2 730	—	2 730	100.0
Vietnam, North	129	—	129	100.0	380	—	380	100.0	450	—	450	100.0
Totals	13 182	26 513	(13 331)		38 086	82 616	(44 530)		(35 079)	70 130	(35 051)	
Regional deficit (surplus)			(13 331)				(44 530)				(35 051)	

Source: U.N., *World Energy Supplies, 1950-1974.* Statistical Series J, No. 19. Tables 10 and 6.

Table 1.4 shows an estimate of the future cost of oil imports in the current net importing countries, should their dependence on imported sources continue at 1974 percentage levels. Projections of oil consumption for 1980 and 1985 are based on energy demand elasticities used in 1976 by the World Bank.[25] Even if output in 1985 from the five producing countries were to remain at 1974 levels, imports for the region would be less than the region's surplus. Again, individual countries with inadequate or no production would still have to take care of their oil requirements with imports, and thus have to balance their foreign exchange oil import costs against their foreign exchange earnings. Even at projected consumption levels more conservative than those presented in Singapore at the Offshore Technology Conference in February 1976,[26] two countries will still face import bills of about US$1 billion each by 1985; these are the Philippines and Thailand.[27]

Oil as a Revenue Earner

In countries where petroleum has been discovered in generous amounts, petroleum has become a major source of domestic revenues as well as foreign exchange. In Indonesia, government revenue from the petroleum industry rose from 5 per cent of total domestic revenue in 1966 to 39 per cent in the fiscal year 1972-3. In FY1974-5 revenues from the petroleum industry amounted to 55 per cent of total revenues.[28] In the international accounts, gross foreign exchange earnings from petroleum in FY1974-5 amounted to 72 per cent of the total, compared to 50 per cent in FY1972-3 and to only 30 per cent in 1966.[29]

Malaysia's petroleum industry, at this stage, contributes a fraction to total foreign exchange earnings. It may not form as significant a source of foreign exchange as Indonesia's industry, mainly because it has other major important export commodities such as palm oil, rubber and tin. Nevertheless, in 1974 its petroleum industry exports grossed close to US$300 million.[30]

About 90 per cent of Brunei's domestic revenues comes from the oil and gas industry, most of which is exported. At the end of 1973, its international reserves amounted to US$279 million; by the end

TABLE 1.4
CONSUMPTION AND IMPORTS OF PETROLEUM PRODUCTS, ACTUAL (1974) AND PROJECTED (1980 AND 1985)

| | Consumption (10³m.t.) | | | Imports | | | | Actual crude oil output (10³m.t.) |
| | Actual | Projected^a | | 10³ m.t. @ 1974 ratios^b | | 10⁶ US$ @ US$11.75/bbl. (1976 dollars) | | |
	1974	1980 @ 4.4% p.a.	1985 @ 4.0% p.a.	1980	1985	1980	1985	1974
GROUP A								
Brunei	71	90	110	—	—	—	—	9 284
Indonesia	7 893	10 200	12 400	—	—	—	—	67 979
Malaysia	3 968	5 130	6 240	—	—	—	—	4 000^c
GROUP B								
Burma	992	1 280	1 550	135	165	11.63	14.21	888
Cambodia	85	110	130	110	130	9.47	11.20	0
Laos	129	165	200	165	200	14.20	17.23	0
Philippines	7 953	10 270	12 490	10 270	12 490	884.53	1 075.73	0
Singapore	3 091	4 000	4 860	4 000	4 860	344.51	418.58	0
Thailand	7 737	10 010	12 170	10 000	12 155	861.28	1 046.88	10
Vietnam, South	2 730	3 530	4 290	3 530	4 290	304.03	369.49	0
Vietnam, North	450	580	700	580	700	49.95	60.29	0
Totals	35 079	45 365	55 140	28 740	34 990	2 479.60	3 013.61	82 161

Source: 1974 consumption and production data from United Nations, *World Energy Supplies*, Series J. No. 19, Tables 10 and 6. Growth rates based on World Bank estimates in *Price Prospects for Major Primary Commodities*, Report No. 814/76, June 1976 (Washington, D.C.: World Bank).

a Based on 1972-1980, and 1981-1985 energy demand elasticities of .874 and .814, respectively, estimated by the World Bank, GNP growth rates of 5% and 4.9% respectively, and the relationship % Δ energy consumption = energy demand elasticity.

$$\frac{\%\ \Delta\ \text{energy consumption}}{\%\ \Delta\ \text{GNP}} = \text{energy demand elasticity.}$$

b Ratios of imports to total supply.

c Estimated. U.N. data do not include Jerudong output in Sabah.

N.B. Projections are rounded downwards.

of 1976 unofficial reports indicated these reserves were expected to exceed US$1.2 billion.[31]

Urgency for Developing Indigenous Resources

Mainly because of the need to meet its consumption requirements but also with the hope that petroleum could provide net foreign exchange gains, South-East Asian countries are under pressure to turn toward exploration and development of their indigenous petroleum resources, despite a supply surplus in world markets. Each country hopes to increase foreign exchange earnings from exports or to reduce the foreign exchange costs of imports. The push from the second factor increases with the prospect of increasing consumption of imported oil for the purpose of economic development.

1.   This ratio may be referred to as the income elasticity of demand for energy. The elasticity concept is a device to indicate the degree of responsiveness of one variable to changes in another variable. In this case, the ratio shows the relationship between relative changes in energy consumption per capita resulting from relative changes in real GNP per capita. In symbolic form,

$$e_y = \frac{\Delta y/y}{\Delta x/x} \text{ or } \frac{\% \,\Delta y}{\% \,\Delta x}$$

2.   The ratio for South Vietnam was negative but may be considered an aberration in view of the abnormal conditions obtaining in that country during that period. Energy consumption reported for that country no doubt included consumption for non-productive uses in the usual economic sense.

3.   The Philippines and Cambodia had zero growths in energy per capita and GNP per capita, respectively.

4.   See Tables 10 and 11 in Lido Gonzalo, 'Philippine energy: supply and demand situation', draft of PREPF Research Note No. 56, May 1976 (Manila: Development Academy of the Philippines).

5.   There is, of course, an indirect relationship between transport modes and productivity. The time saved by driving a personal car may be used by that person for productive activity, i.e., activity that would add to national output, or he may use it for leisure activity that would have no effect on the GNP. If too many people do this, however, energy consumption for transport use may accelerate at rates faster than those of other sectors or of the GNP.

6.   It is not completely eliminated because of, among other things, sectoral differential growth rates that will not be captured in aggregated data.

7. See United Nations, Department of Economic and Social Affairs, table published in 'World energy requirements and resources in the year 2000', *Peaceful Uses
of Atomic Energy*, 1972. In this model GDP per capita was the sole independent
variable. The basic model was of the form:

$$\log C_p = a + b \log \text{GDP}_p + e,\ \text{where}\ C_p = \text{per capita consumption of}$$

energy in kilograms, GDP_p = per capita gross domestic product, and e = a
stochastic variable.

8. See Annex II, pp. 10-12, World Bank, *Energy and Petroleum in Non-OPEC
Developing Countries*, 1974-1980, Staff Working Paper No. 229, February 1976.
The model included price effects and consumption lags. The estimating equation
was in the form:

$$\log C_{k_i t} = -a + b \log \text{GNP}_{k_i t} - c \log P_t + d \log C_{k_i t} - 1$$

where C = oil consumption in barrels per capita, GNP = gross national product
in 1972 dollars per capita, P = f.o.b. price of Saudi Arabian light oil, and k_i =
subscript for higher income, middle income and lower income countries. The co-
efficients b and c are income and price elasticities, respectively, of the demand for
oil.

9. Paul A. Samuelson, *Foundations of Economic Analysis* (Cambridge, Mass.:
Harvard University Press, 1947), p. 125.

10. The study by the World Bank of non-OPEC developing countries yielded
price elasticities of − 0.05 to 0.109 using 1950-1973 data.

11. Computed from Table 1a in Organization for Economic Co-Operation a⸳'
Development, *Energy Policy* (Paris, 1963).

12. Alirio A. Parra, 'Some considerations on the demand and supply of petroleum
in the seventies in developing countries', paper given at Interregional Seminar on
Petroleum Refining in Developing Countries, New Delhi, 22 January − 3 February
1973, p. 3.

13. See S.F. Jefferson, 'Evolution and changes in the pattern of oil supply and
demand in the ECAFE region including Australia and New Zealand', in United
Nations, *Proceedings of the Second Symposium on the Development of Petroleum
Resources of Asia and the Far East*, Vol. II (New York, 1963), pp. 235-40, Table
189.

14. Parra, op. cit., p. 7 and OECD, op. cit., p. 49.

15. In Thailand, bagasse, a by-product of sugar production, is used for producing
steam for operating machinery in sugar mills. Bagasse output increased from 2.2
million metric tons in 1973 to 2.9 million metric tons in 1974, representing 6.9 per
cent of total energy production. Paddy husk for steam production amounted to
257 000 metric tons. Fuel wood and charcoal are also still largely used. Fuel wood
in 1974 amounted to over 1 million cubic metres, and charcoal to over 300 000 cubic
metres. See Thai Delegation, Committee on Natural Resources, United Nations
ESCAP, 'Progress in energy development in Thailand', paper presented at Second
Session, 14-20 October 1975 (Bangkok), pp. 9-10.

In Malaysia charcoal from wood has provided a portion of the energy fuel for both industry and residential consumers. In 1970 about 20 million cubic feet of wood were converted to charcoal. With the discovery of oil in Malaysia, the projected increase in charcoal consumption in Malaysia may not occur. See Peninsular Malaysia, Department of Forestry, 'Energy, resources and the environment — a Malaysian forest management perspective', paper presented at Fifth International Symposium on Energy, Resources, and the Environment, Penang, Malaysia, 21-23 February 1975 (W/ERE/WP/10).

16. In the U.S. in 1974 every barrel of crude run consumed the equivalent of 1/12 barrel of oil. See U.S. Bureau of Mines, *Mineral Industry Survey*, 'Crude petroleum, petroleum products, and natural gas liquids: 1974 (Final Summary)', Petroleum Statement, Annual, prepared 1 April 1976, Table 22, last column.

17. As defined in U.N. statistics, the Far East developing countries, also referred to as 'less developed countries' (LDCs), include, besides the ten South-East Asian countries, Bangladesh, Sri Lanka, India and Korea.

18. See Table 2, *World Energy Supplies*, 1950-1974, and Table 1, 'Energy resources and energy development trends in the ECAFE region', in *Proceedings of the Twelfth Session of the Sub-Committee on Energy Resources and Electric Power*, p. 82.

19. World Bank, *Price Prospects for Major Primary Commodities*, Report No. 814/76 June 1976 (Washington, D.C.).

20. Although Malaysia was a net importer in 1974 this country is expected to be a net exporter as soon as its proved reserves are brought into full production and with further development of its potential. Malaysia is, therefore, excluded from the group of net importers.

21. During the first eleven months of 1975, nearly one-third of exports of crude and partly refined crude from Sarawak were reported to have gone to Japan. *Petroleum News S.E.A.*, August 1976, p. 20.

22. *PN*, August 1976, p. 20. The refining industry in Singapore was reported to be in the process of re-designing its facilities to accommodate lighter crude from the region. See *Asian Wall Street Journal*, 17 February 1977, pp. 1, 7, 'Singapore's economy is entering a phase of reduced growth'.

23. *Bangkok Post* (Thailand), 16 March 1976, 'Burma's oil output increases by 2.6%', p. 10.

24. *Far Eastern Economic Review*, 25 June 1976, 'Tapping Burma's onshore oil', p. 84. The authenticity of this projection is not verifiable.

25. World Bank, *Price Prospects for Major Primary Commodities*, Annex IV, p. 2.

26. See Table 1 in P.D. Gaffney *et. al.*, Economic appraisal of the potential petroleum resources of the Asian Pacific retgion', Offshore Southeast Asia Conference, 19 February 1976.

27. The estimate for the Philippines is also considerably more conservative than that estimated by the Development Academy of the Philippines (DAP) of over 20 million metric tons consumption in 1985. The DAP estimated an energy/GNP elasticity of 1.56 or almost double that used by the World Bank. See Gonzalo, op. cit.

In mid-1977 there were reliable reports that the Philippines would have a small oil producing field off Palawan. Development of Thailand's small gas field (4.4 trillion cubic feet reserves) in the Gulf of Thailand also appeared imminent. While this would constitute some foreign exchange savings, this would only slightly reduce the total oil import bill.

28. U.S. Embassy, *Indonesia Petroleum Report*, 1975 (Jakarta: June 1975). p. 6.

29. Ibid., p. 4.

30. Bank Negara Malaysia, *Quarterly Economic Bulletin*, March/June 1975.

31. *PN*, August 1976, pp. 12 and 14.

II

The Off-shore Petroleum Resource Potential of South-East Asia

INTEREST in the potential of South-East Asia's petroleum resources dates back to over a century. Drilling for oil began in 1872 in West Java,[1] and production began in 1893.[2] The first well in Burma was drilled in 1887.[3] The Miri field in Sarawak was discovered in 1910,[4] and production began in 1911.[5] Exploration for oil in Brunei began in 1911 in what is now known as the Belait anticline, and oil was discovered in the Seria field in 1929.[6] Initial attempts to find oil in the Philippines in commercial quantities were made in 1896 but no commercial discovery has been made as this is written.[7] Interest shifted to the off-shore resources of East Asia in the 1960s, and exploration activity intensified in the late 1960s and early 1970s, involving millions of dollars and thousands of people. This flurry of activity in South-East Asia's seabed may be attributed to technological and economic factors.

Off-shore Resource Estimates

GENERAL ESTIMATES

The economic resources of the continental margin[8] include marine life, minerals, and petroleum resources. By far the most valuable is the last group mentioned—oil, gas, and natural gas liquids. In off-shore areas the most favourable areas for hydrocarbon accumulation in the seabed are thought to be in the sedimentary strata underlying the continental shelf[9] and the upper part of the continental slope[10] (see Fig. 1). Not much is known as yet about the potential of the lower parts of the continental slope. Shallow seas, such as the North Sea and East China Sea, and small ocean basins such as the South China Sea and the Gulf of Mexico, are thought to have relatively high potentials for hydrocarbon accumulation. The existence

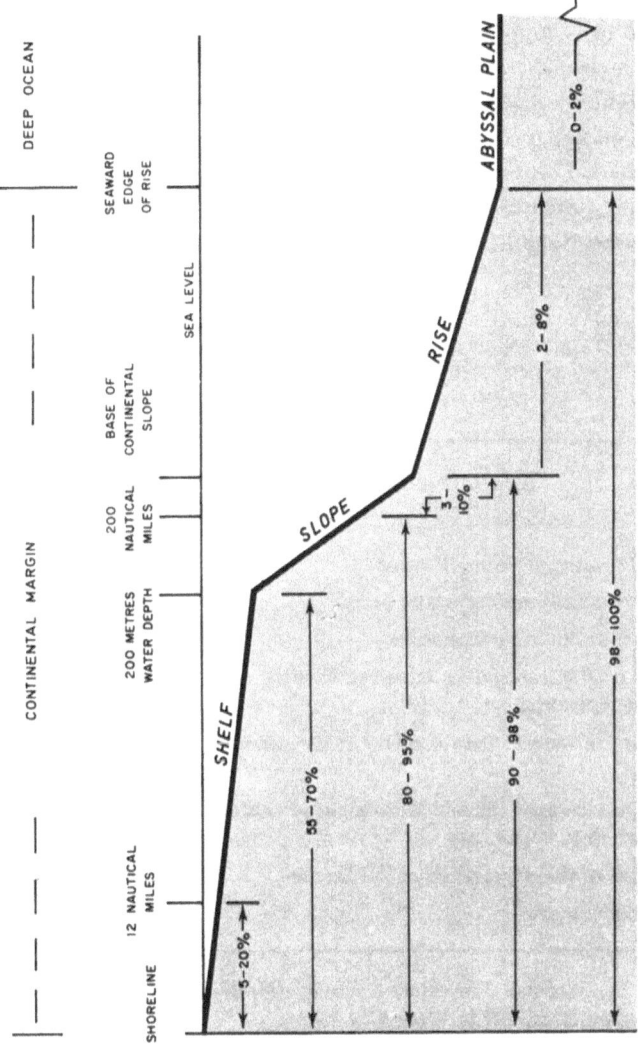

Fig. 1. Estimated Range of World Percentage Distribution of Potential, Ultimately Recoverable
Petroleum within Various Off-shore Boundaries

Source. Adapted from U.S. National Petroleum Council. *Ocean Petroleum Resources,* 1975, p. 17.

of commercially exploitable oil accumulations beyond the continental rise[11] or below a depth of 3 000 metres is considered unlikely.

Table 2.1 shows the various proposed territorial or economic zone boundaries discussed in the Third Law of the Sea Conference. It also shows that up to 70 per cent of total recoverable hydrocarbons in the seabed is expected to be found on the world's continental shelves to the 200-metre depth, and that as much as 95 per cent of that recoverable oil and gas would be located in the 200 nautical mile proposed economic zone. A total of over 2 000 billion barrels of proved reserves and potential resources were estimated to be located within the 200-metre isobath.[12]

TABLE 2.1
ESTIMATED RANGE OF DISTRIBUTION OF POTENTIAL, ULTIMATELY RECOVERABLE PETROLEUM WITHIN VARIOUS OFF-SHORE BOUNDARIES

Off-shore area	World percentage distribution
Within 12 nautical miles of shore	5–20
Shoreward of 200 metres water depth	55–70
Shoreward of 200 nautical miles	80–95
Seaward of 200 nautical miles to the base of continental slope	3–10
Shoreward of base of the continental (or insular) slope	90–98
Between the base of the continental slope and the seaward edge of the rise	2–8
Shoreward of the seaward edge of the rise	98–100
Seaward of the rise	0–2

Source: U.S., National Petroleum Council, *Ocean Petroleum Resources* (Washington, D.C., 1975), Table 3, p. 16.

Metric equivalents:
    1 nautical mile = 1.852 kilometres
    12 nautical miles = 22.224 kilometres
    200 nautical miles = 370.4 kilometres

ESTIMATE OF SOUTH-EAST ASIA'S RESOURCES

The off-shore petroleum potential in South-East Asia is indicated in the following summary:

> From Singapore to Seoul, the greater part of the coastal area of the Asian continent geologically has no potential for oil.... The coastline from Singapore to Seoul, however, is bordered by a continental shelf covering more than a million square miles, within which the broad reconnaissance surveys made to date have already proved the existence of extensive deep sedimentary basins with attractive petroleum potential.[13]

Table 2.2 gives a tabulated description of the coastal geography of the countries of South-East Asia, the ranking of their potential, recoverable petroleum resources, and current estimates of proved reserves of recoverable oil and gas. It will be noted that about one-fifth of the continental shelf area of the world to the 200-metre isobath—that part of the underwater area where hydrocarbon accumulations are considered likely to exist—are in South-East Asia. In fact, looking at a contoured map of South-East Asia, a viewer will note that the islands of Java, Sumatra, and Borneo to the south and the Malaysia-Thailand and Indo-chinese peninsulas to the north appear to be sitting on one broad shelf. The area of Indonesia's continental shelf to the 200-metre depth is, in fact, slightly larger than its total land area of about 2 million square kilometres[14] and accounts for one-eighth of world total shelf area.

Estimated off-shore oil reserves in three currently producing South-East Asian countries now account for about 23 per cent of total world off-shore reserves. These amount to only about 18 per cent of the total proved (on-shore and off-shore) reserves in the region (see last three columns of Table 2.2). The latter reserves are far below the estimated Far East potential (see Table 2.3).

Factors Enhancing South-East Asia's Petroleum Prospects

Three major factors may be cited as enhancing the attractiveness of South-East Asia as a potential petroleum producer. First, advances in off-shore technology allowed oil companies to manifest concrete interest in potentially attractive off-shore areas that had

TABLE 2.2
OFF-SHORE GEOGRAPHY AND POTENTIAL OIL RESOURCES OF SOUTH-EAST ASIA

Country	Coastal category[a]	Coastal length (n. mi.)	Continental shelf to 200-metre depth (sq. n. mi.)	Continental margin (Sq. n. mi.) to 3 000-metre depth	Continental margin to 200 n. mi.	U.S.G.S. report Off-shore Oil	Off-shore Gas	Total Oil	Total Gas	Off-shore Oil (10⁶ bbls.) 1/1/75	Estimated proved reserves Total 1/1/76 Oil (10⁶ bbls.)	Gas (10⁹ cu. ft.)
BRUNEI	Open shelf	88	2 800	5 300	7 100	D	E	D	D	2 287[c]	2 000.0	8 700
BURMA	Open shelf	1 230	66 900	111 300	148 600	D	C	C	C	.0	70.0	160
INDONESIA	Archipelago	19 784	809 600	1 229 800	1 577 300	C	C	C	B	702[d]	14 000.0	15 000
KHMER REPUBLIC	Shelf-locked	210	16 200	16 200	16 200	D	D	C	D	0	0	0
LAOS	Land-locked	0	0	0	0	—	D	—	D	0	0	0
MALAYSIA	Open shelf	1 853	108 900	125 600	138 700	C	C	C	C	364[e]	2 500.0	15 000
PHILIPPINES	Archipelago	6 997	52 000	65 000	551 400	C	C	C	C	0	0	0
SINGAPORE	Shelf-locked	28	100	100	100	E	E	E	E	0	0	0
THAILAND	Open shelf	1 299	75 100	94 700	94 700	D	D	D	D	0	0.2	0[f]
VIETNAM NORTH	Shelf-locked	382	22 200	22 200	22 200	E	E	E	E	0	0	0
VIETNAM SOUTH	Open shelf	865	95 600	151 400	188 400	D	D	D	D	0	0	0
TOTAL S.E.A.		32 736	1 249 400	1 821 600	2 744 700					3 353	18 570	23 860
TOTAL WORLD		181 379	6 467 000	13 738 200	25 279 300					14 683	658 686	2 232 122

Source: John Albers *et al., Summary Petroleum and Selected Mineral Statistics for 120 Countries, Including Offshore,* U.S. Geological Survey Professional Paper 817 (Washington, D.C.: U.S. Government Printing Office, 1973); Sherwood Frezon, *Summary of 1972 Oil and Gas Statistics for Onshore and Offshore Areas of 151 Countries,* U.S. Geological Survey Professional Paper 885 (Washington, D.C.: U.S Government Printing Office, 1974); *International Petroleum Encyclopedia,* 1975; *Oil and Gas Journal,* 29 December 1975.

a See Appendix A.

b Notation Legend.   (10⁹ bbls. of oil or 10¹² cu. ft. of gas.)

    A = 1 000 – 10 000   D = 1 – 10

    B = 100 – 1 000   E = 0.1 – 1

    C = 10 – 100

c Published reserves tend to be conservative. For example, a private report shows Fairley field & Bakau field to be 10 times higher than published.

d Does not include Nora, Zelda, Melahin, and Sepinggan fields discovered in 1974 and 1975, which may contain at least 200 million bbls.

e Covers only major fields of Bakau, Baram, Baronia and West Lutong. No published breakdown available for Samarang and Tembungo.

f As of 31 December 1976, Thailand's gas reserves were reported to be trillion cubic feet (*Oil and Gas Journal,* 27 December 1976). By September 1977 these reserves were reported to be as much as 4.4 trillion cubic feet (*Asian Wall Street Journal,* 1 September 1977).

TABLE 2.3

ESTIMATED CRUDE OIL RESOURCES OF THE WORLD,
BY REGION, 1 JANUARY 1974

Region	Total ultimately recoverable resources (10^9 barrels)	Cumulative production and proved + probable resources, as of 1.1.74 as % of total resources (%)	Estimated off-shore undiscovered recoverable oil as % of total (%)
U.S.A.	230	70.0	20.4
CANADA	85	16.5	68.2
NORTH SEA	50	40.0	60.0
OTHER WEST EUROPE	19	21.1	63.2
MIDDLE EAST	630	79.2	4.3
NORTH AFRICA	84	54.8	9.5
GULF OF GUINEA	52	51.9	38.5
OTHER AFRICA	27	14.8	66.7
N.W. LATIN AMERICA	120	55.8	7.5
OTHER LATIN AMERICA	54	29 6	48.1
FAR EAST[a]	129	27.9	57.4
ANTARCTICA	20	0	100 0
COMMUNIST COUNTRIES	500	30.0	5.8
WORLD TOTAL	2 000	51.9	18.9

Source: Developed from Table III, in J.D. Moody, 'An Estimate of the World's Recoverable Crude Oil Resources', *Proceedings of World Petroleum Congress* (Tokyo, 1975).

[a] Excluding Communist countries.

hitherto been inaccessible. Off-shore drilling by Shell resulted in a discovery off Brunei in 1963. An extensive seismic survey of East Asia's shelves was later conducted under the auspices of the United Nations' Committee for the Coordination of Joint Prospecting in Asian Off-shore Areas (UNDP/CCOP); the report indicated the presence of economically recoverable accumulations of hydrocarbon in the region.[15] A second important factor was the acceptance of the concept of production-sharing that evolved in Indonesia during the years 1966 to 1969, which introduced a new dimension to petroleum development by foreign companies.[16] A third factor was the acceleration in crude oil prices which further changed the cost/price relationship for off-shore oil in the Far East.

Any assessment of an area's potential for oil and gas accumulations is thus essentially dependent on technology and cost/price factors. Previous knowledge about the area, often gained only from actual drilling and the success or failure of others, is also important in a subsequent stage. Even when made against such background assessment is, at best, approximate. The size and type of petroleum deposits vary greatly. The quality of the petroleum, the proportions of oil and gas, and the geologic factors controlling each reservoir differ. Advances in exploration and production technology have improved prediction and recovery factors. Price factors have also changed the dimension of the field over which exploration groups can try their luck. Still, all the uncertainty surrounding what lies beneath the surface of the earth, and in what form, makes it extremely difficult to predict, first, where accumulations will be found and, second, what the ultimately recoverable amounts are.

ADVANCES IN TECHNOLOGY

Perhaps a classic example of how changing technology and economic conditions can alter the petroleum potential of a country can be found in the Malaysian case. In 1959 a U.N. report included the following evaluation of the petroleum prospects in the then Federation of Malaya (now Peninsular Malaysia):

Although a large proportion of the area of the Malay Peninsula has not been mapped geologically in detail, it is evident from what is already known that the prospects of finding large workable reserves of petroleum at depth to be very remote. Although the Federation of Malaya is surrounded by the petroleum producing countries of Burma in the north, Sumatra to the west, and Borneo to the east, analysis of the tectonic trends shows that ... the oil-producing areas fall basically into two major zones lying on either side of this tin-producing region.

. . . .

A recent comprehensive aerial magnetometric and scintillo-metric survey covering approximately 16,000 square miles, or nearly one-third of the area of the Federation of Malaya, has revealed no indication to [sic] the accumulation of petroleum.[17]

In February 1976 an article on the future prospects of off-shore oil ranked countries by output potential. Malaysia was ranked ninth in terms of potential output of about 47 million tons per year from the South China Sea, Sarawak and Sabah. Indonesia ranked 12th, and Brunei 13th.[18]

Most of the earlier off-shore exploration in the region was conducted in marine areas adjacent to producing land areas. In recent years oil exploration has been directed to the marine shelves where very little was known about the prospects of production, but where similar exploration has been rewarding in other parts of the world. The recent discoveries in Indonesia along the Australian North West Shelf as well as in Malaysia and Brunei in the South China Sea and in the Gulf of Thailand have provided evidence that the petroleum potential of South-East Asia is substantially greater than previously believed. Production is now concentrated on the shelf surrounding Borneo, but the oil and gas prospects of the western coast of the Thai-Malay peninsula — that is, the Straits of Malacca area and its extension into the Andaman Sea — are thought to be far from negligible.[19]

Exploration drilling in north-west Borneo's continental shelf started in 1956 at water depths of less than 200 metres, and in 1957, Asia's first off-shore well was drilled on this shelf.[20] In 1973 Mobil signed a production-sharing contract for an area with an average water depth of 1 000 metres in the Macassar Strait (between the islands of Sulawesi and Borneo).[21]

In 1975 the UNDP/CCOP noted that exploration had progressed increasingly into the deeper waters of the continental slopes of the region. The CCOP reported that more acreage covering deep-water areas had been contracted in South-East Asia than in any other region, and this was attributed to the fact that the geologic regime and ocean environment were considered particularly favourable and attractive for deep-water exploration and exploitation.[22]

Dramatic changes in technology have taken place in the last ten years that have already improved exploration and estimation techniques.[23] Although the deeper waters of South-East Asia are largely untested, exploration in the near future may be expected to cover the small basins, trenches and outer parts of the continental slopes.

CHANGES IN COST/PRICE STRUCTURE

The upward shift in the world price structure of oil in the 1970s has also made off-shore production commercially viable and South-East Asia's resources potentially attractive. In 1972 most of the world's off-shore production came from areas less than 30 metres deep (about 100 feet). At that time it was thought that since development costs increased with greater drilling depths, values of US$5.00-7.00 per barrel were necessary to make deep-water production commercially feasible.[24] The current price, no doubt, has improved the commercial prospects of certain geologically attractive areas like the North Sea and South-East Asia.

Another economic aspect of off-shore production is the size of what is considered a commercial prospect. Because of higher operating costs, an off-shore well must produce several times as much as one on-shore to be considered commercially attractive. An off-shore rig of average complexity is reported to cost around US$20 000 to US$35 000 per day to operate, depending on both technological and economic conditions. A rig operating off Burma in March 1976 was reported to have been hired by an oil company at a cost of US$20 000 per day; including all other expenses, contracted drilling operations cost around US$40 000 per day.[25] On some of the more sophisticated drillships, these costs run up to US$90 000 per day.[26] Thus, an off-shore exploratory well in East Asia was expected to cost an average of around US$5 million. In the earlier years of off-shore exploration, oil firms were said to require that, to be commercially viable, a South-East Asian off-shore well should produce a minimum of 2 500 barrels per day in a field capable of a total production of about 50 000 barrels per day—in contrast to a minimum of 500 barrels per day for a commercial field on-shore. Multiple price increases in the 1970s may suggest that fields capable of producing 25 000 barrels per day and per-well production of 1 250-1 500 barrels per day (or less) could prove to be commercial.[27]

A third economic factor that reportedly made South-East Asia's resources attractive in the early 1970s, given some of its relatively less attractive geological aspects, was acquisition cost. A group of foreign oil companies acquired more than 140,000 square miles (360 000 square kilometres) of off-shore concessions in Indonesia in the early 1970s for a signature bonus or initial cash outlay of less than US$2 million.[28] In the United States, oil leases on about 850 square miles (2 200 square kilometres) of continental shelf cost oil companies US$845.8 million in initial payments in November 1970.[29] The financial requirements for new tracts in South-East Asia have risen since then, but the inter-regional cost-relationships have not changed drastically.

Off-shore Production

The share of off-shore oil in total oil supply in South-East Asia has increased rapidly over the past ten years. Off-shore production has grown at an annual average of about 30 per cent between 1965 and 1975. At its post-war peak in 1973, off-shore production amounted to 505 million metric tons and accounted for 14 per cent of world production.[30] Estimates of 1975 off-shore production are still incomplete at this writing, but Table 2.4 shows that, even with incomplete estimates, off-shore oil still amounts to about 14 per cent of total output.

In 1965 the region produced only 28 million metric tons of crude oil; ten years later it was producing over 70 million tons, over one-fifth of which was from off-shore wells. Output rose at an average rate of over 100 per cent between 1971 and 1975 in Indonesia, and at half that rate in Malaysia.

Table 2.5 shows that by 1980 off-shore oil could account for over

TABLE 2.4

OFF-SHORE PRODUCTION OF PETROLEUM IN SOUTH-EAST ASIA, 1965, 1975

	Natural gas liquids (10⁶ metric tons)			Crude oil						Natural gas total (10⁶ cu. metres)		
				Total (10⁶ metric tons)			Off-shore (10⁶ metric tons)					
	1965	1975	Rate of growth (%)	1965	1975	Rate of growth (%)	1965	1975	Rate of growth (%)	1965	1975	Rate of growth (%)
WORLD	50.990	89.952	5.8	1 509.52	2 646.50	5.8	30.86	370.0^a	28.5	675.68	1 268.84	6.5
FAR EAST	0.061	0.087	3.6	32.73	88.72	10.5*	4.75 (1970)	N.A.	48.5(4)	7.56	26.88	13.5
BRUNEI	0.057	0.062	0.9	3.94	9.5	8.7	3.8 (1970)	7.0^c	12.8(5)	0.211	6.00	40.0
BURMA	—	—	—	0.54	1.03	6.6	—	—	—	0.008	0.015	6.5
INDONESIA	0.015 (1970)	0.005	-19.5	23.95	64.46	10.3	0.54 (1971)	12.24^d	123.0(4)	3.16	6.06	6.7
MALAYSIA	—	—	—	0.05	4.70^b	57.5	0.8 (1971)	4.0^c	50.0(4)	—	—	—
THAILAND	—	—	—	0.002	0.01	17.4	—	—	—	—	—	—

Source: United Nations, World Energy Supplies, Stat. Series J. No. 19 (New York, 1976), Tables 6 and 15; Offshore Magazine, 24 June 1976, p. 7; Oil and Gas Journal, 29 December 1975, and 3 May 1976; U.S. Bureau of Mines, Mineral Industry Surveys, 'World Crude Oil Production. Annual, 1975'; issued 9 June 1976, and World Natural Gas; 1974', issued 23 June 1976; Socialist Republic of the Union of Burma. Report to the Pyithu Hluttaw on Financial, Economic, and Social Conditions for 1975-76 (Rangoon: Ministry of Planning and Finance).

* Not adjusted to show changes below.

^{a} Own estimate developed from Oil and Gas Journal, 3 May 1976, p. 150, and adding U.N. off-shore data for Centrally Planned Economies for 1974.

^{b} Own estimate based on data appearing in Oil and Gas Journal, 29 December 1976, and U.S. Bureau of Mines, Mineral Industry Survey, 'World Crude Oil Production. Annual, 1975'. The U.N. estimate for Malaysia was only for Sarawak and was lower by around 400 000 metric tons; for Brunei the U.N. estimate was over 500 000 metric tons higher than OGJ and USBM estimates. Conversion factors were based on USBM factors for the countries concerned (Malaysia: 7.709 bbl. = 1 m.t.).

^{c} Oil and Gas Journal, 3 May 1976, p. 148. Conversion factor 7.334 bbl. = 1 m.t. Also, consistent with Offshore Magazine, 24 June 1976, p. 71.

^{d} Offshore Magazine, 24 June 1976, p. 71.

^{e} Own estimate obtained by extrapolating the average in Oil and Gas Journal, 29 December 1975, for Tembungo field in Sabah and adding this to U.N. estimate for 1974. This is about 238 000 metric tons above the total for 1975 shown for 3 major Malaysian fields in Oil and Gas Journal, 3 March 1976, p. 148. This was also found to be consistent with data in Offshore Magazine, 20 June 1975, p. 71.

Note: Numbers in parenthesis under rate of growth denote years of computation less than 10.

24 per cent of total world output. It also shows that over half of pre-
dicted output in the Far East and Australia would be coming from
off-shore wells. As resources on land are depleted, and as technology
for producing in deeper waters advances, the bulk of future output
can be expected to be produced from off-shore fields.

Table 2.6 shows the forecast output of oil and gas for the three
South-East Asian countries currently producing oil in marine areas.
According to these projections, off-shore oil output alone from these
three countries in 1980 will be higher than the total output in 1975 in
all of South-East Asia, and off-shore output from Brunei and Malaysia in 1980 will exceed the 1975 total output of Indonesia, which
accounted for over 80 per cent of the region's oil production.

TABLE 2.5

OFF-SHORE OIL: FORECAST PRODUCTION POTENTIAL BY 1980 * (BY REGION)
(million barrels a day)

Region	Off-shore	Total	Off-shore as % of total
WEST EUROPE	5.07	5.35	94.8
NORTH AMERICA	2·84	14.99	18·9
CENTRAL AMERICA & WEST INDIES	0.51	1.36	37.5
SOUTH AMERICA	0.32	5.43	5.9
AFRICA	2.10	8.32	25·2
MIDDLE EAST	6.65	31·20	21·3
FAR EAST & AUSTRALIA	2.39	4 19	57 0
CENTRALLY PLAN- NED ECONOMIES	2.04	18·92	10.8
TOTAL	21·92	89·76	24.4

Source: Based on Table B of Scottish Council Survey presented in *Petroleum Economist*, February 1976, p. 49.

* High Forecast Case —
    Assumes (1) all announced plans will be carried out.
           (2) excess capacity of around 24% (90 MMBD total production against 72 MMBD demand).

Restrictions: Limited to 200-metre water depth.

TABLE 2.6

OFF-SHORE OIL AND GAS: POTENTIAL OUTPUT IN 1980
(SOUTH-EAST ASIA)
(million metric tons per year)

Country	Off-shore Area	10^6 Tons/ Year
Brunei	South China Sea	22
Indonesia	Java & Ceram Seas, Macassar and Malacca Straits	27
Malaysia	South China Sea, Sabah and Sarawak Coasts	47
Sub-total		96
World Total		1 420
3 countries as percentage of world total		6.8%

Source: Based on Scottish Council survey presented in Table II, *Petroleum Economist*, February 1976, p. 50.

The Scottish Council estimates are, of course, 'high forecasts', which assume (1) that 'all announced plans' will be carried out and (2) that world supply would exceed projected demand by about 20 million barrels per day. Recent developments around Asia may prove these forecasts to be extremely optimistic. These estimates, nevertheless, indicate the growing attractiveness of oil resources beyond the shores of each country.

The survey said nothing of other South-East Asian countries which have no proved commercial deposits beyond their shores. This does not preclude the possibility that Burma, Thailand, the Philippines or Vietnam may join this group of countries, as exploration continues in the waters of the South China Sea, the Gulf of Thailand, and the Andaman Sea. There have been oil and gas shows in wells drilled on the shelves of these countries. Of special interest here are those in the Gulf of Thailand and in the Spratly Archipelago. These are discussed in later chapters in connection with actual and potential ownership disputes.

1.   Soembarjono, Ministry of Mines, Indonesia, 'Petroleum offshore activities and availability of natural gas in Indonesia', in United Nations, ECAFE, CCOP, *Report of the Sixth Session,* 1969 (Bangkok), p. 113.

2.   U.S. Embassy, *Indonesia Petroleum Report, 1973/74* (Jakarta: July 1974), p. 3.

3.   'Status and methods of petroleum exploration in Burma', in United Nations, ECAFE, *Proceedings of the Symposium on the Development of Petroleum Resources of Asia and the Far East,* Mineral Resources Development Series No. 10 (New York: 1959), p. 98.

4.   Ng Shui Meng, *The Oil System in Southeast Asia* (Singapore: Institute of Southeast Asian Studies, 1974), p. 17.

5.   F.W. Roe, 'British territories in Borneo', in United Nations, ECAFE, *Proceedings of the Symposium on the Development of Petroleum Resources of Asia and the Far East* (New York: 1959), p. 81.

6.   M.S. Michie, 'The search for oil in Brunei', in *Petroleum di-Brunei* (Brunei Shell Petroleum Company, April 1975).

7.   Cora Siddayao, 'Looking for oil in the Philippines', *Esso Eastern Review* (New York City: June 1966), p. 3. See footnote 27 of Chapter 1.

8.   The term *continental margin* is the geological term used to refer to the prolongation of the continental land mass into the ocean (see Fig. 1). It is composed of the *continental shelf,* the *continental slope,* and the *continental rise.* Beyond the continental margin is what is referred to as the *abyssal plain.* Definitions of the latter four will be made as they first appear subsequent to this. Unless otherwise footnoted, definitions are from three sources: S.E. Frezon, U.S. Geological Survey, *Summary, 1972 Oil and Gas Statistics for Onshore and Offshore Areas, 151 Countries,* Professional Paper 885 (Washington, D.C.: Government Printing Office, 1974), pp. 2-3; Evan Luard, *The Control of the Sea-Bed* (London: William Heinemann, Ltd., 1974), pp. 29-30; and John Temple Swing, 'Who will own the oceans?', *Foreign Affairs,* April 1976, pp. 528-46.

9.   Geologically, the *continental shelf* is defined as the area between the shoreline and the point in the sub-water extension of the land mass where the inclination becomes perceptibly greater. The average gradient of the shelf is about 0.1 degree and the kink occurs usually at a depth of from 130 to 200 metres; in exceptional cases the kink occurs at a point as shallow as 50 metres or as deep as 500 metres. Its width may range from about 1.6 to 1 300 kilometres (1 to 800 miles) from the shoreline.

10. The term *continental slope* refers to the steeper area adjacent to the edge of the continental shelf up to the point where the inclination becomes steeper; its edge occurs at depths of about 1 500-2 500 metres, and its width is relatively narrower than that of the shelf (about 15 to 30 kilometres, depending on the cut-off point).

11. The term *continental rise* refers to the generally steeper area between the continental slope and the point where it merges with the true ocean bottom or the ocean bed. The U.S. Geological Survey defines this outer limit in terms of the 3 000-metre isobath, although other authors write that this can occur at a depth of 5 000 metres. The width of the rise is estimated to be between 160 and 1 600 kilometres.

12. 'Economic significance, in terms of seabed mineral resources, of the various limits proposed for national jurisdiction', Report of the Secretary-General of the United Nations dated 4 June 1973, for submission to the Committee on the Peaceful Uses of Sea-Bed for Ocean Floor Beyond the Limits of National Jurisdiction (Doc. A/AC. 138/87), partially reproduced in Table 20-I, in the U.N., ESCAP, CCOP *Report of the Tenth Session*, 1973, p. 186.

In this paper the term 'resources' is distinguished from the term 'reserves'. The term 'resources' refers to concentrations of petroleum discovered, undiscovered, and surmised to exist in such form that extraction is currently or potentially feasible. The term 'reserves' refers to that subset of resources that not only already have been identified by geological or engineering methods but can also be produced at a profit and can be legally extracted at the time of reporting. The distinction is important, because while oil resources as a whole are non-renewable depletable resources, proved reserves may increase or decrease depending on the discovery of new locations of accumulations of oil resources and development of such discoveries. See Corazón M. Siddayao, 'Patterns in the utilization of the major energy resources of the United States', report written for the Ford Foundation, June 1974, pp. 3 and 4, partially reproduced in 'Appendix D—major energy resources', *A Time to Choose: America's Energy Future* (Ballinger Publishing Co., 1974), pp. 477-84.

13. M. Mainguy, 'Regional geology and petroleum prospect of the marine shelves of eastern Asia', in CCOP, *Technical Bulletin*, No. 3, May 1970, p. 103.

14. See K. Romimohtarto, A.G. Ilahude, and A. Notji, 'Marine research in Indonesia with special reference to the National Institute of Oceanology (L.O.N.)', in *Report of the Ninth Session*, 1972, of the ECAFE, Committee for Coordination of Joint Prospecting in Mineral Resources in Asian Waters (CCOP), Bangkok, 1973, p. 76.

15. See United Nations, ECAFE, CCOP, *Technical Bulletin*, Volume 2, issued May 1969, and Volume 3, issued May 1970.

16. See A.G. Hatley, 'Offshore petroleum exploration in East Asia—an overview', *Proceedings of Offshore Southeast Asia Conference*, SEAPEX Programme Paper I (Singapore, 1976).

17. 'The prospects for petroleum in the Federation of Malaya', in United Nations, ECAFE, *Proceedings of the Symposium on the Development of Petroleum Resources of Asia and the Far East* (Bangkok, 1959).

18. Ranking by Scottish Council Survey, Table II, in 'Over 20 million b/d in the 1980's?', *Petroleum Economist*, February 1976, pp. 49-51.

19. Mainguy, op. cit., p. 92.

20. *Petroleum News S.E.A.*, July 1973, p. 40.

21. Ibid., p. 34.

22. UN, ESCAP, CCOP, *Report of the Twelfth Session*, 1975, p. 6.

23. Some of these are: (1) sophisticated echo sounder which give a visual profile of the ocean floor; (2) seismic and sonic profiling of the rocks and deposits beneath the seabed; (3) gravimetric equipment moved along the sea floor with remote control, which allows interpretation of the structures and types of rocks beneath the seabed;

(4) aircraft and satellites which can measure sea temperature currents, ice, sea level, pollution and other features; (5) aerial photography which can penetrate clear, shallow water to record the sea bottom, as well as under-water cameras and television which give pictures of the bottom of the sea, showing the distribution of rocks, nodules, etc., and (6) new methods of core drilling which enable large samples to be taken from considerable depth. See Luard, op. cit., pp. 23-4.

24. CCOP, *Report of the Tenth Session*, 1973, p. 185.

25. These figures were obtained by the author from a drilling superintendent in Burma, March 1976. The US$40 000 figure does not include salaries of oil company personnel monitoring the drilling operation. Total company cost of drilling in Thai waters was reported to exceed US$100 000 per day. See *Petroleum News S.E.A.*, March 1977, p. 7.

26. Hatley, op. cit.

27. 'Searching for Oil and Gas in Asia', in Leon Howell and Michael Morrow, *Asia, Oil Politics and the Energy Crisis* (New York: IDOC/North America, 1974), p. 60.

28. Cal-Asiatic-Topco agreement, see *Petroleum News S.E.A. Annual*, 1976, p. 17, and Sritua Arief, *The Indonesian Petroleum Industry* (Jakarta: Sritua Arief Associates, 1976), p. 518.

29. Howell and Morrow, op. cit., p. 60.

30. Many sources speak of a 20 per cent share, but this is obviously based on 1972 production when off-shore output accounted for close to 18 per cent of total output. See UN, *World Energy Supplies*, Stat. Series J, No. 19, p. 193.

PART TWO

Potential Areas of Conflict

III
Potential Sources of Conflict and Law of the Sea Issues

THE presence of petroleum resources in South-East Asia's seabed and the increasing value attached to them may give rise to serious international conflicts in the course of the development of these resources. Such conflicts would essentially arise from disagreements over ownership rights. Questions concerning property rights over off-shore petroleum resources could spring from (1) incompatibility in perceptions of the equity of existing or proposed legal definitions of jurisdictional boundaries concerned, and (2) incompatibility of jurisdictional boundaries so delineated with geological and environmental phenomena. Disputes may also arise from conflicting historical ownership claims. As stated in the introductory chapter, this study will attempt to identify such conflict situations only for the purpose of analysing their economic and policy implications. No attempt will be made to delve into the politics of such situations.

Conflicts or disagreements in the course of developing off-shore petroleum may arise in all phases of the industry: in exploration; in development and production; and in transporting, refining, and marketing the oil produced. Disagreements with regard to property rights may be encountered as early as the exploration stage, since the right to explore is limited by jurisdictional boundaries. Environmental concerns may, however, give rise to problems in the exploratory stage when seismic work and exploratory drilling are conducted, and property rights certainly will be involved in problems that could result from the non-excludable nature of geophysical and environmental conditions in the production, transportation, and marketing phases. It is thus not feasible to categorize problems by industry phases. Instead the discussion will be broken down according to the source of conflict.

Jurisdictional Issues in the Law of the Sea Affecting Off-shore Petroleum Resources

Ownership rights over seabed petroleum resources were legally redefined, along with redefinition of certain jurisdictional boundaries, by conventions adopted at the First Conference on the Law of the Sea held in 1958 in Geneva. Since 1958, however, the definitions have become either inadequate or meaningless. Many coastal states have extended jurisdiction over economic resources as far as possible seaward, as changes in technology and economic conditions increased the abilities of such states to do so. Pressures to redefine juridical boundaries yet another time led to the Third Conference which began in 1973.

The issues at the Law of the Sea Conference that bear most directly on South-East Asia's off-shore petroleum resources are those of the economic zone and the continental shelf; to a lesser extent, the issues of territorial limits, pollution control, and scientific research are also relevant. The interplay of conflicting national interests has delayed resolution of those issues, and some of the difficulties will be discussed below. Before discussing these issues and South-East Asian national differences, and to appreciate the complexity of property rights disagreements over marine resources, it would be appropriate to give a short background to the whole sea law issue.

CHANGES IN THE LIMITS OF THE JURISDICTIONAL SEA

The traditional fundamental doctrine of the law of the sea is 'freedom of the seas'. This concept was propounded by the Dutch jurist, Hugo Grotius, in his *Mare Liberum* ('free sea') of 1605, in opposition to the dogma of *mare clausum* ('closed sea') proposed by an English scholar, John Selden.[1] Grotius wrote, 'The sea, since it is as incapable of being seized as the air, cannot have been attached to the possession of any particular nation'.[2]

Under international law as it evolved through the centuries, the ocean was divided into *territorial waters* (defined as the marine area adjacent to the coast of a state where it exercises sovereignty over the sea, air space, seabed, and subsoil) and the *high seas* (defined as the parts of the sea beyond territorial waters, not subject to jurisdiction

by any state). Until a few decades ago, the territorial waters of a
country were, with a few exceptions, limited to the waters within 3
miles of the shoreline.[3] Originally developed in relation to national
security, the limit was established in the eighteenth century when
the effective range of a cannon ball from a warship was 3 miles.

Extension of territorial waters beyond the traditional 3 miles became common after the First World War, but claims to even wider
limits became more frequent after the Second World War. As the
potential of discovering oil beyond the 3-mile limit became technologically achievable both in fact for the shallower parts and in
prospect for the deeper parts, the United States declared in the Truman Proclamation of 1945 that it regarded the 'natural resources
of the subsoil and seabed of the continental shelf beneath the high
seas but contiguous to the coasts of the United States as appertaining
to the United States, subject to its jurisdiction and control'.[4] It re-
served to the United States the right to explore and exploit these
resources.

Other Latin American countries similarly made unilateral de-
clarations, claiming national jurisdiction over areas extending 200
nautical miles from the shore.[5] These countries had narrow continental margins, and therefore limited potential oil and gas resources;
but they had fertile fishing grounds off their coasts, and were increasingly threatened by technologically advanced fishing fleets from
the United States.[6]

The wider territorial claims of the Latin American countries
were followed by other nations. Brunei annexed its continental shelf
by proclamation in June 1954.[7] The Philippines claimed historic,
archipelagic waters in 1955,[8] and Indonesia claimed archipelagic
waters in 1957.[9] By 1975 only 29 of the 149 coastal states recognized
the 3-mile territorial limit, with over a third claiming 12-mile limits
and at least eight claiming 200-mile territorial seas.[10]

U.N. CONFERENCES ON THE LAW OF THE SEA

The First United Nations Conference on the Law of the Sea in
1958 adopted a Convention setting a 12-mile maximum on the territorial sea but failed to set a common limit. The Convention also provided for a contiguous zone for customs, fiscal, immigration, or sani-

tary regulations purposes not to exceed 12 miles from the territorial sea baseline. The First Conference also produced, in addition to the Convention on the Territorial Sea and the Contiguous Zone, a convention that codified freedom of the high seas, and two conventions relative to exploitation of living resources, namely, a convention on fisheries and another on the continental shelf.

The Second U.N. Conference on the Law of the Sea in 1960 failed to arrive at universal agreement on the width of the territorial sea. In the face of rapid technological and scientific change that threatened national security, on the one hand, and which enhanced the prospects of augmenting national wealth, on the other, more and increasingly frequent claims extended the national waters of coastal nations and jurisdiction over ocean resources beyond their original limits.

The question of territorial waters brought up a host of other issues. They included (1) the archipelagic issue; (2) the definition of an island; (3) the right of innocent passage; (4) the right of passage through heretofore international straits; (5) the outer limit of an exclusive economic zone or national authority over ocean resources; (6) the question of closed or semi-closed seas; (7) the right of states to prevent pollution; and (8) the right of states to regulate marine research.

All issues have arisen from the fact that governments have tried to protect 'national interests', defined subjectively. As Luard stated: 'They accept the need for mutually acceptable arrangements. But their first aim is to ensure that the arrangements are acceptable to themselves.'[11] The different issues all sum up to concern about national property rights.

In his now famous speech of 1 November 1967 before the United Nations, Ambassador Arvid Pardo of Malta called for a declaration of the seabed and the ocean floor beyond 'present' national limits[12] as 'a common heritage of mankind' which should be 'used and exploited for peaceful purposes and for the exclusive benefit of mankind as a whole'.[13] This came as claims by coastal states reduced other nations' rights to this common ownership.

In 1970 the U.N. General Assembly voted to convene a third Conference on the Law of the Sea to deal with a range of issues. The

Third Conference met for 2 weeks in New York in December 1973, 10 weeks in Caracas from June to August 1974, 8 weeks in Geneva in March and April 1975, 8 weeks in New York in March and April 1976, and 8 weeks again in New York in August 1976. Over 150 countries were participating. The fifth session ended in a deadlock. A sixth session was held from May to July 1977 in New York; while some gains were made, no agreement was in sight, and a seventh session was scheduled for March 1978.

A document called the Single Negotiating Text was produced in the 1975 Geneva session. The March-April 1976 New York session produced a Revised Negotiating Text (397 articles and 11 annexes) which was taken up at the August session. By the end of the latter session, there appeared to be a broad consensus on a 200-n. mile exclusive economic zone and extension of territorial waters from 3 to 12 nautical miles. The Revised Text recognized that the resources of the seabed beyond national jurisdiction were 'the common heritage of mankind'. It also provided that exploitation should be administered by an international regime, which would either do the mining itself or contract it out to private companies under its supervision. No agreement was in sight on this latter problem.

A new topic was raised on the mechanism for settling disputes arising from a Law of the Sea Treaty,[14] and the fourth part of the Revised Text for use in the sixth session in mid-1977 was devoted to this.[15] At the end of that session, there was no agreement on the issues involved.[16]

A new document to serve as the basis for future negotiations was produced at the end of the sixth session. Called the 'Informal Composite Negotiating Text',[17] this document incorporated the changes negotiated at the preceding sessions.

The Territorial Sea Issue

The width of the territorial sea was not defined in the 1958 Geneva Convention on the Territorial Sea and the Contiguous Zone. The maximum limit of 12 miles was implied from the definition of the contiguous zone as follows:

TABLE 3.1

NATIONAL POSITIONS ON LAW OF THE SEA ISSUES

Country/Territorial and related issues	Resource development	Research	Pollution
BRUNEI			
	– Continental shelf annexed June 1954. (Proclamation, 30 June 1954.)		
BURMA			
– Territorial sea: 12 nautical miles from baseline. (Declaration, 15 November 1968); reaffirmed by Law No. 3 of 1977, promulgated 9 April 1977.)	– 1958 Convention definition of the continental shelf supported. (See Concession Rules of 1962, cited in *Limits in the Seas*, p. 34.)	Prior consent of state is mandatory for any research conducted on and about the continental shelf. (*Official Records*, II, p. 155.)	Exclusive jurisdiction proclaimed in territorial sea, continental shelf and economic zone. (Law No. 3 of 1977.)
– Single island, rock, or group of islands or rocks has independent territorial sea extending 12 nautical miles from low-water line on its coasts. (Ibid.)	– Seabed is 'common heritage of mankind' and an international regime to regulate development of international seabed area should be set up. Benefits to be derived from such exploitation were to be distributed among all countries of the world. (*Official Records*, II, p. 26.)	Legal requirements for prior express permission to conduct scientific research on territorial sea, continental shelf or economic zone embodied in Law No. 3 of 1977.	
– Contiguous zone: 24 nautical miles. (Law No. 3 of 1977.)	– Continental shelf regime is autonomous regime within the broader framework of an exclusive economic zone or patrimonial sea; the natural prolongation principle is supported; coastal states have total jurisdiction over continental shelf, including living resources. (Ibid., p. 155.)		
	– Sovereign rights proclaimed on 9 April 1977 over living and non-living resources of continental shelf, defined as submarine areas (seabed and subsoil) extending beyond the territorial sea throughout the natural prolongation of its land territory to the outer edge of the continental margin, or to a distance of 200 nautical miles from baselines where shelf does not extend up to that distance. (Law No. 3 of 1977.)		

CAMBODIA

- 12 nautical miles of territorial sea. (Declaration, 27 September 1969.)
- Party to 1958 Convention.

- Economic zone declared covering outer limit of 200 nautical miles from territorial sea baselines; jurisdiction over both living and mineral resources. (Law No. 3 of 1977.)

- Claimed full sovereignty over the continental shelf and ownership of, and therefore control and jurisdiction over the natural resources on the shelf. (Declaration, 27 September 1969.)
- Party to 1958 Convention on Continental Shelf.
- Boundaries delimited in Decree No. 439-72. PRK. 1 July 1972.

LAOS

- Archipelagic principle expounded by Indonesia and the Philippines is supported. (Official Records, II, p. 272.)

INDONESIA

- Archipelagic principle. (Official Records, III, p. 226.)
- 12 nautical mile limit for territorial sea, determined from the baselines applicable to archipelagic states. (Ibid., II, p. 114. See also Act No. 4, 18 February 1960. Document A/Conf. 19/5/Add. 1, 4 April 1960.)

- Jurisdiction proclaimed to whatever depth resources can be exploited. (Government announcement, 17 February 1969, cited in Limits in the Seas, 36, p. 95.)
- Delimitation of continental shelf to 200-metre isobath rejected, because many countries have already concluded agreements covering limits extending beyond it; would combine concept of continental shelf and 200-mile zone. (Official Records, II, p. 169.)

Sovereign rights over research within in economic zone. (Ibid., II, pp. 207-8, 268.)

Sovereign rights over protection of marine environment with economic zone. (Ibid., pp. 207-8.)

TABLE 3.1 (*Continued*)

Country/Territorial and related issues	Resource development	Research	Pollution
Re reservations of Malaysia concerning adverse effect of archipelagic principle for access and communication between West and East Malaysia. Indonesia believes inclusion of the provision of non-interference in the direct communication between parts of the territory of an adjacent state is sufficient. (Ibid., II, pp. 292-3.) — Jointly recommended with Malaysia revision in Composite Text that would preserve rights traditionally exercised or stipulated in agreements covering archipelagic waters of a state, where these waters lie between parts of an immediately adjacent neighbouring state. (Private source.)	— Supports concepts of economic zone and patrimonial sea. (Ibid., p. 207.) — Ownership proclaimed over all mineral and other living and non-living resources in the seabed and subsoil outside territorial limits to where superjacent waters admits its exploitation and undertaking. Where such shelves have a border with another state, the border line would be determined by negotiation. (Proclamation concerning the Indonesian Shelf, 1 Feb. 1969, submitted as Document CCOP (IV)/4.)		
MALAYSIA — 12 nautical mile territorial sea. (See *Emergency (Essential Powers) Ordinance*, No. 7, 2 Aug. 1969, cited in *Limits in the Seas*, p. 122.) — Reservations held about archipelagic principle, in so far as Indonesian archipelagic boundary would affect free access and communication between West and East Malaysia. (*Official Records*, II, p. 198.) — Nevertheless, official position was to support archipelagic concept despite disadvantage for	— Continental shelf defined as the seabed and subsoil of those submarine areas adjacent to the territorial limits of Malaysia to a depth no greater than 200 metres below the surface of the sea, or to depth where superjacent waters admit of exploitation of the natural resources, living and non-living. (See Continental Shelf Act, 28 July 1966.) — Party to 1958 Convention. — Establishment of an exclusive economic zone beyond the territorial sea, up to a limit of 200 nautical miles. Juris-	Jurisdiction over scientific research in economic zone. (Ibid., II, p. 198.)	Jurisdiction over marine environment in economic zone. (Ibid., II, p. 198.)

itself in the interest of friendship, on condition that the legitimate and existing rights of other states adversely affected by this principle would be protected by international law. (Ibid., II, pp. 292-3; see also section on Indonesia above.)

– Sought revision of provision in Revised Text (Article 42, Part II) that denied coastal state right to regulate passage through straits. (Straits Times, 14 May 1977; see also section on Indonesia above.)

PHILIPPINES

– Territorial sea is in the form of two inverted triangles, according to treaty limits defined in the Treaty of Paris, 1898, between Spain and the United States ('historic waters'). (Republic Act No. 3046, 17 June 1961.)

– Above limits are at their widest, 270 miles toward the Pacific and some 145 miles toward the China Sea, and at narrowest in the south, less than 3 miles. (See Tolentino, Phil. Geog. Journal, Vol. XIX, 4 Nov. 1975, p. 158.)

– Archipelagic principle, first proposed to UN

– Jurisdiction over exploration and development of minerals and all other living and non-living resources on the continental shelf. (Proclamation No. 370 of 20 March 1958.)

– Resources of the sea are a common heritage of mankind and concept of exclusive economic zone supported. (Tolentino, p. 158.)

– International regime for exploitation of seabed resources. (Ibid., II, p. 23.)

diction in the zone would be limited to living and non-living resources.

Reservations held about extent of economic zone applicable to archipelagic states.

Concept of continental shelf subsumed in that of an exclusive economic zone was rejected. (Official Records, II, pp. 198, 278.)

– Rights over continental shelf already acquired under existing international law should continue to be recognized. (Ibid., I, p. 144.)

– Establishment of international regime to regulate development of international seabed beyond the limits of national jurisdiction, and the sharing of revenues acquired therefrom. (Ibid., I, pp. 144-5.)

– Jurisdiction over research within archipelagic waters. (Document A/Conf. 62/C.2/L.49, Official Records, III, pp. 226-7.)

– Protection of marine environment through international regime. (Doc. A/Conf. 62/C.3/L.6, Official Records, III, p. 240.)

– States should have right of control in economic zone.

TABLE 3.1 (*Continued*)

Country/Territorial and related issues	Resource development	Research	Pollution
by RP in 1955 under concept of historic waters, is also supported. RP refused to sign 1958 Geneva Conventions because Law of the Sea Conference had not adopted the archipelagic principle. (*Official Records*, II, p. 264.) — Philippines proposed that the maximum limit proposed in the Convention on Territorial Seas should not apply to historic waters held by any state as its territorial sea. (*Official Records*, III, p. 202.)			Zonal approach to preservation of marine environment proposed. (Doc. A/Conf. C.2/C.3/L.6.)
SINGAPORE — 12-mile limit. (*Official Records*, II, p. 211, and Doc. A/Conf. 62/C.2/L.33.) — Concept of archipelago accepted for Indonesia and Philippines provided the 'legitimate interests and rights' of the international community and neighbours are taken into consideration, e.g. fish caught by Singaporeans in what is at present considered 'high seas' but which would become Indonesian waters under the archipelagic principle. (Ibid., II, p. 268.)	— 'Doctrine of creeping jurisdiction' and division of the oceans into exclusive 370-kilometre economic zones for coastal states opposed. Concept of jurisdiction over continental shelf resources (non-living) opposed when interpreted to extend to any distance beyond shores. Extension of economic zone to continental margin or 3 000-metre isobath opposed as this would leave to GDS no area where hydrocarbon accumulations are likely to happen; negates concept of common heritage of mankind. (*Official Records*, II, p. 151.) — Part of resources in ocean floor should be under an International Seabed Authority to allow land-locked and	Control of scientific research over economic zone. (*Straits Times*, 14 July 1975.)	

— GDS nations to share (common heritage of mankind concept). (See *Straits Times*, 15 March 1976.)

— Notion of equating 12-mile territorial limit to economic zone, with area beyond limit to be under International Seabed Authority, supported. International Seabed Authority would exploit non-living resources.

  Regional approach for such exploitation supported, that is, economic zones should be regional or sub-regional. (Ibid., II, p. 211; I, p. 134.)

— Non-renewable resources of the economic zone should be governed by the principle of the 'common heritage of mankind'; revenue-sharing system should be adopted. (Ibid., I, p. 135.)

— Island states, like Singapore, should be entitled to an economic zone; however, not every island should have an economic zone (e.g., uninhabited, small). (Ibid., II, p. 285.)

— Party to 1958 Convention on the Continental Shelf.

— Position of GDS countries supported. (*Official Records*, II, p. 22.)

— Common heritage of mankind concept supported, and establishment of international regime with broad authority over management and development of ocean resources. (*Official Records*, II, p. 22.)

— Retention of the concept of the continental shelf favoured

— Coastal states should have broad competence to ensure their own security and the bonafide nature of research in waters.

International standards in marine control of pollution. (*Official Records*, I, p. 148.)

THAILAND

— Principle of archipelagic states favoured in general but interests of neighbouring states that may be affected by new law should be considered, e.g., access to sea and fishing resources. (*Official Records*, I, p. 147.)

— 12 nautical mile territorial sea. (Royal Proclamation, 6 Oct. 1966, cited in *Limits in the Seas*, p. 192; claim repeated in UN Document I/NR/R.91.) (Ibid., I, p. 148.)

TABLE 3.1 (Continued)

Country/Territorial and related issues	Resource development	Research	Pollution
	or any new international law that would not adversely affect agreement already reached between countries on boundaries of the shelf.	– Research benefits should be shared. (*Official Records*, I, p. 188.)	
	No logical or legal reason was seen for the exclusion of the continental shelf concept from the concept of the economic zone or patrimonial sea.		
	1958 Geneva Convention's definition was considered defective in that the concept of exploitability made it open-ended. (Ibid., II, p. 159.)		
	– 200-mile economic zone acceptable on condition of 'satisfactory' solution to what comes under a state's jurisdiction and on compensation to countries that do not have the ability to extend their jurisdiction, such as sharing living resources on an 'equitable basis' with other states. (Ibid., p. 192.)		
VIETNAM, NORTH			
– 12 nautical mile territorial sea limit. (Proclamation, 1 Sept. 1964, cited in *Limits in the Seas*, p. 221.)			
VIETNAM, SOUTH			
– 3 nautical mile territorial sea declared on 27 April 1965. (Decree No. 81, cited in *Limits in the Seas*, p. 222.) Territorial waters reported extended from 12 to 50 nautical miles in 1974. (*Petroleum News*, 31 May 1974, p. 5.)	– Subsoil and seabed and all natural resources of continental shelf belong to the Rep. of Vietnam. (Proclamation, 7 Sept. 1967.)		
	– Shelf defined to 200-metre depth and exploitability.		

– Rights over continental shelf will not affect legal status of the 'superficial' waters as high seas or of airspace above waters. (Proclamation. 7 September 1967.)

(11/70 Law, 1 December 1970, cited in *Limits in the Seas*, p. 222.)

– Shelf area delimited by Decree No. 249/BKT/VPU-GQGDH/ND. 6 Sept. 1971. (*Limits in the Seas*, p. 222.)

– 200-mile limit for economic zone acceptable except where shelf extends beyond that distance, in which case limit would be at the edge of continental margin. (*Official Records*, II, p. 162.)

– Applicability of economic zone concept to mid-ocean islands regardless of size questioned. (Ibid., II, pp. 192-3.)

– Coastal states' exclusive use of non-living resources in economic zone considered equitable. (Ibid., II, p. 255.)

VIETNAM, UNIFIED
– Jurisdiction claimed over 12 nautical mile territorial sea plus 12 nautical mile contiguous zone, declared 12 May 1977: applies to islands and archipelagos. (Vietnam News Agency and U.S. State Department.)

– Jurisdiction claimed over 200 nautical mile economic zone and over continental shelf throughout natural prolongation to margin or 200 nautical miles (whichever distance is greater), proclaimed on 12 May 1977. (Ibid.)

Source: Except where otherwise cited above, positions are taken from United Nations, *Third Conference on the Law of the Sea, Official Records,* Vols. I-III, VI, 1973-1976 sessions (New York: 1975 and 1977); and from U.S. State Department, *Limits in the Seas Series,* No. 36: *National Claims to Maritime Jurisdictions,* 3rd Revision, December 1975. Additional citations include developments subsequent to 1974 sessions.

1. *In a zone of the high sea contiguous to its territorial sea,* the coastal state may exercise the control necessary to:
   (a) Prevent infringement of its customs, fiscal, immigration or sanitary regulations within its territory or territorial sea; . . . .
2. The contiguous zone *may not extend beyond twelve miles from the baseline* from which the breadth of the territorial sea is measured.[18] [Italics added]

The problem of a common outer limit for the territorial sea was complicated by the introduction of the archipelagic principle by Indonesia and the Philippines. The 12-mile territorial sea concept also raised problems over the status of straits—sea lanes that were once considered international waterways but which would, under the new limit, fall under the jurisdiction of coastal states.

In South-East Asia, only Malaysia and Thailand were parties to the Convention.

SOUTH-EAST ASIAN COUNTRIES' VIEWS ON TERRITORIAL LIMITS

By the start of the fifth session of the Third Conference on the Law of the Sea there was a consensus on an extension of territorial waters to 12 nautical miles and of the contiguous zone to 24 nautical miles, and these were embodied in the Revised Single Negotiating Text under discussion.[19] Almost all South-East Asian countries supported the 12 nautical mile limit[20] (see Table 3.1). South Vietnam was the only one known to have extended its territorial waters to 50 miles, with a Saigon declaration reported in the press in May 1974.[21] Subsequently, in a statement by the Vietnam News Agency dated 12 May 1977 and reported in the foreign press, the new government of unified Vietnam, the Socialist Republic of Vietnam, proclaimed its sovereignty over territorial waters of 12 nautical miles and an adjoining contiguous zone of identical width, or a total maximum width extending 24 nautical miles from the baseline.[22] This, in effect, reaffirmed the 1964 claim of the former North Vietnam government to a 12 nautical mile territorial sea.[23]

In 1968 Burma declared its territorial sea to be 12 nautical miles from the baseline, including that for single islands, rocks, or groups of islands of rocks within its territorial waters.[24] This distance

55

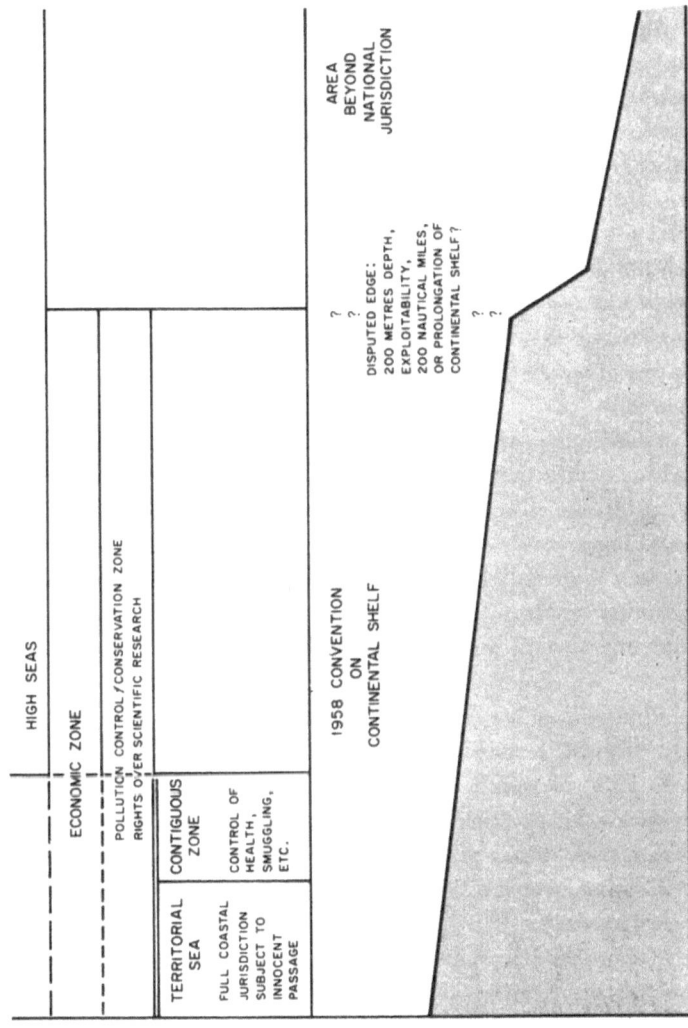

Fig. 2. Boundary Terms in the Law of the Sea
(With acknowledgements to Evan Luard, *Control of the Sea-Bed*)

was reiterated in a law passed by the People's Congress on 28 March 1977; in addition, this law declared a 24 nautical mile contiguous zone.[25] In 1969 Cambodia also claimed a 12 nautical mile territorial sea.[26] Singapore, Laos, and twenty-two other geographically disadvantaged states (GDS)[27] co-authored draft articles specifying a territorial sea limit not exceeding 12 nautical miles.[28]

Indonesia has supported the 12 nautical mile limit only in so far as this would apply to acceptance of the archipelagic principle, that is, in so far as such limit would be measured from the baselines applicable to archipelagic states.[29] Indonesia first made its claims known in a government declaration in 1957, which was codified as Government Law No. 4/1960.[30] Indonesia consists of 13 667 islands, and a 'twelve-mile [sic]' territorial sea would cover 6.8 million square kilometres of water. Its land area consists of only 2 million square kilometres.[31]

Indonesia signed several treaties delimiting its territorial sea. As of the time of this writing it had territorial sea treaties with Malaysia on the Straits of Malacca (ratified 1971), with Singapore on the Straits of Singapore (ratified 1974), and with Australia on behalf of Papua New Guinea (1973).[32] In mid-1976 it was said to be negotiating similar territorial sea agreements with India and Thailand.[33] The agreement with India was reportedly signed on 14 January 1977.[34]

The Philippines has supported the archipelagic principle and claimed 'historic' territorial waters.[35] According to the Treaty of Paris of 1898, the limits of such waters, at their widest, extend 270 miles toward the Pacific and some 145 miles toward the China Sea, and, at their narrowest in the South, are only 3 miles in breadth.[36] The Philippines was indifferent to any maximum territorial limit set by the Convention.[37]

The Philippines has argued that when it became a sovereign state in 1946 it acquired the territory ceded by Spain.[38] The historic waters concept was incorporated in the archipelagic concept embodied in the 'draft articles relating to archipelagic States' proposed by Fiji, Indonesia, Mauritius and the Philippines.[39] This concept was included in the Revised Text as an exception to delimitation procedures.[40] The Philippine archipelago is composed of over 7 000 islands, some of which have a lot of sea between and around them. If the territorial sea of the Philippines were to be defined by 12-mile limits around each island, the Philippines would lose about 600 000 square kilometres (230 000 square miles) of territorial sea.[41]

Land-locked Laos, while supporting the 12 nautical mile territorial limit, also supported the archipelagic principle in regard to Indonesia and the Philippines.[42] But Singapore and Thailand— both GDS—were sympathetic to the archipelagic claims of those two countries only in so far as such principle did not infringe on their rights of access to the high seas and to living resources. Singapore was willing to accept the archipelagic principle provided 'legitimate interests and rights' of the international community and neighbours are taken into consideration.[43] Thailand, because of its location on the 'semi-enclosed' sea,[44] has supported the position of the GDS countries, and sponsored a draft article specifying a requirement to enter into bilateral agreements to allow equal opportunity to exploit the living resources in archipelagic or national waters of a neighbouring country.[45]

Malaysia has claimed a 12 nautical mile territorial sea.[46] It expressed reservations about the archipelagic principle for two main reasons: (1) it would convert high seas under the previous international regime into national waters of archipelagic states, and (2) it would provide states with the opportunity to claim economic zones that it considered would result in an 'inequitable and unbalanced apportionment of marine space'.[47]

Malaysia's reservations were largely concerned with the break in its territorial waters that would result from the application of the archipelagic principle to Indonesia's islands in the South China Sea. It claimed the archipelagic principle would restrict air and sea communications between Peninsular Malaysia and eastern Malaysia.

Indonesia argued that inclusion in the Convention of a provision of non-interference in the direct communication between parts of the territory of adjacent states would solve the problem.[48] In fact, the Negotiating Text provided for this.[49] Moreover, Indonesia did not believe that rights related to an 'accident of geography' which had preceded political unification of the west and east Malaysian territories could be disputed.[50]

Support of the archipelagic principle by the Philippines and Indonesia was based on political and economic reasons: (1) national security and (2) exclusive exploitation of the resources of the waters, seabed and subsoil within these waters. The most crucial issue in the principle is not the breadth of the waters but the baselines for measuring both the continental shelf, as legally defined, and the economic zone.

Jurisdictional Issues over Resources

The most divisive issue in the Third Law of the Sea discussions has been that related to jurisdiction over ocean resources: the continental shelf, the economic zone, and control over the seabed beyond national jurisdiction. The issue of the legal definition of the continental shelf is especially relevant to the exploitation of petroleum resources; however, it cannot be discussed separately from the issue of the economic zone concept, for the second may be considered to have evolved from disputes over rights or non-rights granted by the first.

THE CONTINENTAL SHELF CONCEPT

The 1958 Convention tried to reconcile the economic needs of nations and claims to sovereignty over the waters by adopting the convention on fisheries and that on the continental shelf.

The Convention on the Continental Shelf granted to a coastal state sovereign rights over its shelf to explore and exploit non-living resources of the seabed and subsoil and living organisms that 'are either immobile on or under the seabed or are unable to move except in constant physical contact with the seabed or the subsoil.'[51] The Convention defined the *continental shelf* as referring to the 'seabed and subsoil of the submarine area adjacent to the coast but outside the area of the territorial sea, to a depth of 200 metres or, *beyond that limit, to where the depth of the superjacent waters admits of the exploitation of the natural resources of the said areas.*'[52] In the western Pacific region, only Cambodia, Malaysia, and Thailand ratified the Convention.

In the Composite Text the articles on the continental shelf reflected the variant positions of many coastal states with diverse natural endowments as well as the position taken by the International Court of Justice (ICJ) in 1969 on the boundaries of certain shelves in the North Sea case. The proposed articles defined the continental shelf as comprising the seabed and subsoil of the submarine areas located beyond the state's territorial sea 'throughout the natural prolongation of its land territory to the outer edge of the continental margin, or *to a distance of 200 nautical miles* from the [territorial] baselines ... where the outer edge of the continental margin does not extend up to that distance'.[53] Exploration and exploitation rights similar to those provided in the 1958 Geneva Convention were provided for in the Composite Text.[54] Beyond the 200 nautical mile zone, that is, the area beyond national jurisdiction, the Composite Text provided that any exploitation by the coastal state would require sharing with other nations by payment or contributions in kind to an administrative international authority to be set up for that purpose.[55]

THE ECONOMIC ZONE CONCEPT

The Composite Text stated that the *exclusive economic zone* would be the area beyond and adjacent to a coastal state's territorial sea and limited this zone to a breadth of not more than 200 nautical miles from the territorial baselines.[56] In this zone the coastal state would have sovereign rights over exploration, exploitation, conservation and management of living and non-living natural resources of the bed, subsoil and superjacent waters. It also would have exclusive rights and jurisdiction over (1) production of energy from water, currents, and winds, (2) scientific research, and (3) preservation of the marine environment. The draft articles also provided for a regulation to achieve optimum utilization of the zone's living resources and for access by other nations, including land-locked states.[57] Rocks which cannot sustain human habitation or economic life of their own would not be accorded an exclusive economic zone or continental shelf.[58] The foregoing provisions tried to incorporate the views of many states, including those of the GDS.

COMPLEXITIES SURROUNDING THE TWO CONCEPTS

The delimitation of the continental shelf has been a prime legal issue for many years. With the sharp increase in petroleum prices since 1973, nations started to look at their own oil resources and the settlement of the issue has taken on more urgency. Resolution of the question was not helped by the introduction of the economic zone concept, which took hold in the discussions at the Third Law of the Sea Conference.

The legal definition of the continental shelf under the Geneva Convention combined geological and technological concepts. Both were stated obscurely, leaving the definition open-ended. The definition allowed for variations in the breadth of the jurisdiction over seabed resources, depending on a particular coastal state's natural endowments. In addition, the exploitability criterion further obscured the definition of rights. It contradicted the depth limit, which was at least explicit. What would be the limits of exploitability? Technological developments have extended the ability to exploit further out to sea and to greater depths. Whose ability to exploit would be the criterion—the coastal state's, or technological development in general? In 1975 and 1976 an American oil company drilled in water depths of 1 900 feet (579 m) to 3 461 feet (1 055 m).[59] The oil industry expects capability to improve considerably by 1980 or 1985.

The obscurity of the Convention was deepened with the ICJ decision of 1969, determining the boundaries of some shelves in the North Sea shared by states. The judgment of the Court had been sought when negotiations failed between the Netherlands and the Federal Republic of Germany, on the one hand, and between Denmark and the Federal Republic of Germany, on the other, on delimitation of the further course of their shelf boundaries.[60] On 20 February 1969, the Court ruled, among other things, that:[61]

1. The use of the equidistance method of delimitation was not obligatory on the parties concerned;
2. delimitation, taking into account equitable principles, should leave as much as possible to each party all those parts of the continental shelf that constitute a natural prolongation of its land territory into and under the sea, without encroachment on the natural prolongation of the land territory of the other;[62]
3. such areas which overlap were to be divided in agreed proportions or exploited under a regime of joint jurisdiction.

The Court also ruled that the physical and geological attributes of the continental shelf areas, including the general configuration of the coasts of the parties, should be taken into consideration. The Court's decision crippled the equidistance provision of the Convention.

The juridical definition of the continental shelf proposed in the Revised Text and incorporated in the Composite Text reduced some of the vagueness of the earlier Convention, but not all. It specified the outer limit of 200 nautical miles where the continental margin did not extend to that distance. Yet its inclusion of the term 'natural prolongation' of the land territory meant that the 'sovereign' and 'exclusive' jurisdiction of a state could in fact extend to more than 200 miles, where the shelf exceeded that breadth (which it could). In terms of the 'common heritage of mankind' concept, this still would defeat Arvid Pardo's intent, because it would preempt the ownership rights of other states and would exclude the GDS from the opportunity to share in the exploitation of petroleum resources on the shelf.

Two essential differences remained between the notion of the economic zone and the revised notion of the continental shelf as proposed in the Composite Text. First, the economic zone concept would grant to the coastal state rights over all living resources in the superjacent waters, while the continental shelf concept would not. Second, the economic zone concept would limit sovereignty and jurisdiction to living and non-living resources in the seabed and subsoil only up to a distance of 200 nautical miles from the territorial baseline; the continental shelf concept, on the other hand, would grant to the coastal state rights to whatever distance the shelf extended, but to not less than 200 nautical miles (except for shelves that were shared by adjacent or opposite states).

The issue of the economic zone has been complicated by the dispute over whether or not the rights to resources of the continental shelf granted in the 1958 Convention have become acquired rights which cannot be altered by new legal definitions that may be ratified in the Third Conference on the Law of the Sea. More important, if the 1958 Convention rights have become acquired rights, then a few countries with continental shelves exceeding 200 nautical miles

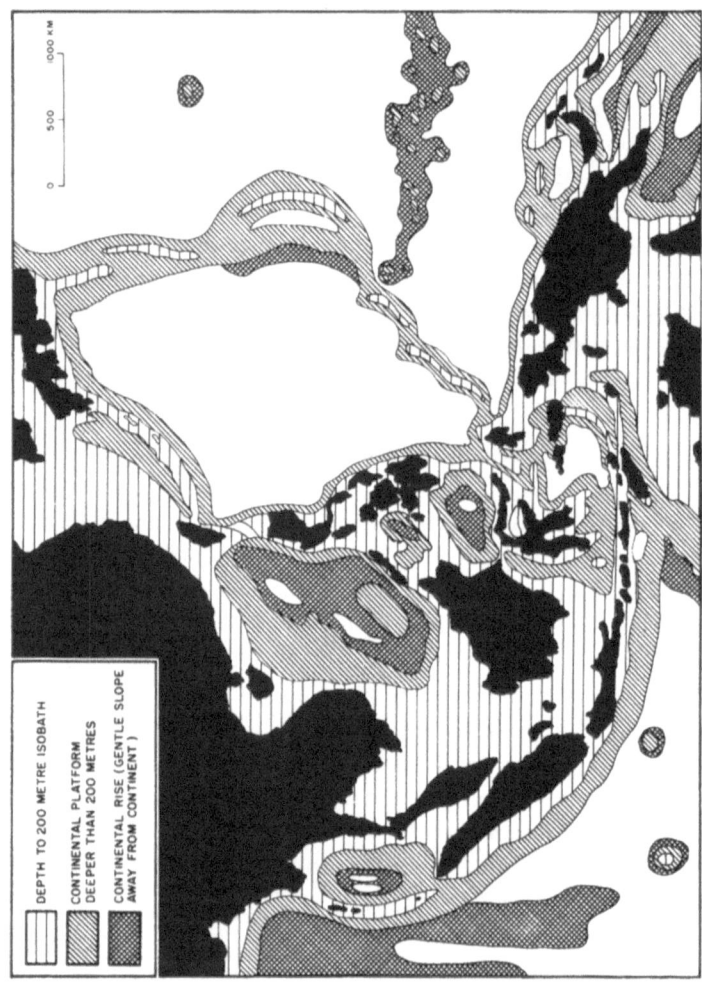

Fig. 3. Seabed Areas in South-East Asia Showing Topographic Gradients

could legally exercise control over vast ocean areas. An estimate showed that three countries—Canada, Australia, and Indonesia—could together control 30 per cent of the world's continental margins to the 3 000-metre depth. Indonesia alone was estimated to have 1 229 700 square miles (3 184 910 square kilometres) and would control 9.21 per cent of these zones.[63] Figure 3 shows the areas that would accrue to coastal states in South-East Asia.

The spectre of a similar take-over of vast areas of the ocean with adoption of the economic zone concept led to a united stand by GDS countries to protect their rights. As early as the preparatory discussions of the Seabed Committee, land-locked countries sought agreement in principle on (1) free access to and from the sea and to the international seabed area beyond national jurisdiction, (2) participation in the international regime, and (3) equitable sharing of the benefits of the area.[64]

Early in the discussions, various limits had been suggested for the exclusive economic zone. These limits were: 200-metre isobath, 500-metre isobath, 40 nautical miles from the shoreline, 200 nautical miles from the baseline, and the edge of the continental margin. In response to the request of Singapore and thirty other GDS countries,[65] the U.N. Sea-Bed Committee studied the economic implications of the various proposed limits.[66]

Luard pointed out that a 200-mile limit would not, in fact, have any genuine basis in geographical reality. He wrote: 'It would not approximate to any known geographical concept. It would mean that coastal nations were claiming huge areas far beyond sight of their own coasts, with which they had no greater contact or relationship than the peoples of other nations. Some countries could claim larger areas of the seabed than they had on land.'[67]

Figure 4 shows how the concept of the continental shelf might be a source of rancour and possible international disagreement where a state is deprived of access to resources of the shelf by an accident of geography or by a political boundary. Where Burma and Thailand share a common boundary on land, Burma's land territory covers a strip along the Andaman Sea much smaller than the corresponding inland area of Thailand directly east of the sea. Yet Burma would have access to all the resources on the prolongation of its narrow strip into the sea. On the other side, Thailand's land area on the coast of the Gulf of Thailand is much smaller than the land area of Cambodia directly east of the Gulf, yet all of the resources of the seabed area would fall within the jurisdiction of Thailand, if boundaries are as drawn.[68]

NON-MUTUAL NATIONAL, ECONOMIC INTERESTS

In the U.N. debates nations have supported either the notion of the economic zone of 200 nautical miles seaward or that of the continental shelf based on the 'natural prolongation' concept, depending on their respective geographical characteristics or natural en-

Fig. 4. Tentative Boundary Lines of Thailand on the Andaman Sea
and Gulf of Thailand
Source: Thailand, Ministry of Industry, Mineral, Fuels, Division,
Petroleum Activities in Thailand, March 1976.

dowments. Nations with extended coastlines have tried to extend their national domain as far out into the sea as possible, while GDS countries have attempted to reduce the area of national jurisdiction and of exclusive economic rights. Thus, among South-East Asian countries differing natural endowments have resulted in positions that reflect such endowments. All countries with broad continental shelves have supported the concept.

Although Indonesia is not listed as having ratified the Convention on the Continental Shelf in February 1969, it proclaimed ownership over all mineral and other living and non-living resources in the seabed and subsoil outside its territorial limits to a depth of 200 metres, or beyond that limit to 'where superjacent waters admit its exploitation and undertaking'. Its proclamation also stated that the borders of shelves adjacent to that of another state would be determined by negotiation.[69] Indonesia subsequently negotiated agreements on the boundaries of continental shelves with Malaysia covering the Straits of Malacca and the South China Sea (1969), with Malaysia and Thailand on the northern Straits of Malacca (1971; ratified 1973), with Thailand (1971; ratified 1973), with Australia (1971; ratified 1973), with India on the Andaman Sea (1974), and with Australia acting on behalf of Papua New Guinea (1973).[70]

Indonesia has strongly objected to delimitation of the continental shelf to the 200-metre isobath. It has done so on the grounds that many countries (including itself) had, pursuant to the 1958 Geneva Convention on the Continental Shelf, concluded agreements with boundaries extending beyond the 200-metre depth. At the Third Conference it proposed combining the concepts of the continental shelf and the 200 nautical mile economic zone.[71]

While supporting establishment of a 200 nautical mile exclusive economic zone beyond the territorial sea, Malaysia has rejected proposals to subsume the concept of the continental shelf in that of the economic zone.[72] Malaysia passed a Continental Shelf Act in 1966.[73] This piece of legislation, like that of Indonesia, codified into its own legal system the concept of the continental shelf embodied in the Geneva Convention, which Malaysia ratified. Malaysia's legislation also defined the term 'natural resources', and provided for exploitation, only under exploration licence or agreement, of petroleum resources on on-shore areas (defined as including the area within its territorial waters) and on contiguous off-shore land (defined as the area of the continental shelf). Having already completed some agreements, Malaysia has supported the continuation of rights over the continental shelf acquired under existing international law.[74]

Burma, Cambodia, Thailand and Vietnam have also supported both the continental shelf and the economic zone concepts (see Table 3.1).

Burma has strongly opposed any proposal that would oblige a coastal state to share with others part of the revenue derived from the exploration of the mineral resources of the continental shelf, or, that would, in effect, dilute ownership over the resources on the shelf. In addition, Burma has insisted that the 'natural prolongation' principle should supersede the economic zone concept where applicable.[75] As exploration for oil off the coasts of Burma began early in 1975, Burma announced that it might 'soon declare a 200' mile economic zone'.[76] Delays in the resolution of the question at the Law of the Sea Conference led the Burmese Foreign Minister to reiterate its plan at the U.N. General Assembly in October 1976,[77] and on 28 March 1977, the People's Congress passed a law extending the country's jurisdiction over living and non-living resources in an exclusive economic zone of 200 nautical miles and over its continental shelf.[78] Promulgated by the State Council on 9 April 1977, the law defined the continental shelf as 'the seabed and subsoil of the submarine areas that extend beyond the territorial sea throughout the natural prolongation of its land territory to the outer edge of the continental margin, or to a distance of 200 nautical miles from the baselines where the outer edge of the continental shelf does not extend up to that distance.'[78]

Back in 1967 Cambodia claimed full sovereignty over the continental shelf and ownership of and, therefore, control and jurisdiction over the natural resources on its shelf.[80]

Thailand has accepted both the concept of the economic zone and that of the continental shelf. At the same time it has supported the positions of other GDS countries in relation to these concepts. It has favoured the adoption of the economic zone concept, provided a 'satisfactory' solution is reached (1) on what comes under a state's jurisdiction and (2) on compensation to countries that are unable to extend their jurisdiction. It has proposed, as a solution, the sharing of living resources on an 'equitable basis' with other states.[81] It has considered equitable the extension of rights over the non-living resources of the economic zone to the coastal state, but has questioned the applicability of the economic zone concept to mid-ocean islands, regardless of size.[82]

Thailand has also favoured the retention of the concept of the continental shelf, although it has acknowledged that the open-ended definition of the 1958 Geneva Convention was defective. It saw no logical reason for excluding the concept of the continental shelf from the concept of the economic zone or patrimonial sea. Having concluded an agreement with Indonesia, it has rejected the adoption of any new international law that would nullify agreements already reached concerning boundaries of the continental shelves between countries.[83]

The government of South Vietnam proclaimed rights over its continental shelf in 1967, although it indicated that such rights 'will not affect the legal status of the superficial waters as high seas or that of the airspace above those waters'.[84] Like Burma, South Vietnam also favoured the adoption of the 200-mile economic zone except where the shelf extended beyond that distance, in which case the limit would be the edge of the continental margin.[85] Where there was bound to be disagreement between nations on such delimitation, it recommended direct, bilateral negotiation or peaceful settlement through an international organization.

The government of reunified Vietnam proclaimed jurisdiction over the resources of the country's adjacent marine areas on 12 May 1977.[86] It claimed rights over an exclusive economic zone of 200 nautical miles and over Vietnam's continental shelf throughout the natural prolongation of the land to the continental margin or to 200 nautical miles where the shelf was narrower than the economic zone.[87] These rights included 'exploring, exploiting, conserving and managing all natural resources'.[88] The announcement was reported to have stated that Vietnam would settle matters relating to the sea zones *through negotiations* and that it would deal with off-shore claims 'in accordance with the principle of defending the sovereignty and interest of the Socialist Republic of Vietnam, and *in keeping with international law and practices*'.[89]

The Philippines, with narrow shelves, has supported the economic zone concept which, together with historic territorial baselines, would grant it vast water areas. At the same time it has supported the notion that exploration and development of living and non-living resources on the continental shelf should be under the jurisdiction of the coastal state. In keeping with the moves of other nations, the Philippines proclaimed in 1968 exclusive jurisdiction and control over the exploration and exploitation of all mineral and other natural resources including living organisms of the sedentary species in the seabed and subsoil of the 'continental shelf adjacent to the Philippines, but outside the area of its territorial sea to where the depth of the superjacent waters admits of the exploitation of such resources'.[90] In fact the Philippines tried to use both the economic zone concept of 200 nautical miles and continental shelf rights to support its rights to drill in the Reed Bank in the Spratly archipelago.[91] The Philippines could not, of course, legally employ the economic zone concept, since it was not a right under existing laws of the sea. In 1977 the Philippines limited its claims to its rights under the Continental Shelf Convention; actually it may not support its claim on physical grounds, as there is a reported deep trench that separates the Reed Bank from the Philippines and which, if present, would end the Philippines shelf well before reaching the Reed Bank. How the Philippines resolves this problem is not within the purview of this study.

Singapore, a geographically disadvantaged country, has opposed what it calls the 'doctrine of creeping jurisdiction'.[92] It has strongly opposed adoption of the economic zone concept, and, more particularly, continuation of the continental shelf concept. Its position on adoption of the exclusive economic zone has hinged on the presence of adequate guarantees in the Convention that land-locked and geographically disadvantaged countries like itself would continue to fish in the waters of zones which previously had been high seas. It has particularly opposed the continuation of the continental shelf concept because this has precluded the exploita ion of all hydrocarbon accumulations on the shelf by GDS nations. It recommended, among other things, (1) equating the 12-mile territorial limit to the economic zone, (2) the development of a regional approach to exploitation of resources, and (3) granting economic zones to GDS countries.[93]

As stated earlier, the Composite Text provided for the extension of rights over the continental margin to at least the breadth granted under the economic zone. It was reported that there was a broad consensus about the adoption of the economic zone concept and, it appeared, the revised continental shelf concept. The fifth session of the Third Law of the Sea Conference ended without resolving these issues mainly because of the failure of the protagonists on the third resource issue — that relating to the seabed beyond these zones — to find a mutually acceptable solution. The point of departure in the positions of developed and developing countries has been whether or not the exploitation of the deep seabed should be limited to states to the exclusion of private firms. Certain developing countries have supported the position that the resources of the deep seas be totally under the control of a mining authority.[94] The sixth session in mid-1977 was not more successful.[95]

An eventual accord on the issues at the Law of the Sea Conference and the signing of a treaty on jurisdictional boundaries do not, however, guarantee an end to disagreements or acrimony between individual countries. That is precisely the reason for a need to adopt procedures on arbitration of disputes, the details of which have not been agreed upon either. Thus, the presence of highly valued petroleum resources in the seabed of South-East Asia remains a potential source of boundary disputes under any regime.

1.   Patrick A. Mulloy, 'Political storm signals over the sea', *Natural History*, 1973. It will be of interest to South-East Asian scholars that the involvement of Grotius in legal maritime issues arose from a Dutch-Portuguese dispute over transit rights in the Straits of Malacca, also an issue in the Third Law of the Sea Conference. Grotius advocated freedom of the seas on behalf of the Dutch East India Company's interests, while Selden defended the right of dominion by Britain over foreign incursions for security and economic reasons. See Seyom Brown and Larry L. Fabian, 'Diplomats at Sea', *Foreign Affairs*, Vol. 52, No. 2 (January 1974), pp. 302-3.

2. Cited in John Temple Swing, 'Who will own the oceans?', *Foreign Affairs*, April 1976, p. 528.

3. Some Scandinavian countries claimed 4 miles, a few Mediterranean countries claimed 6 miles, and Russia in Czarist days claimed 12 miles. See Luard, op. cit., p. 30.

4. U.S.A., Presidential Proclamation No. 2667 on the 'Policy of the United States with respect to the natural resources of the subsoil and seabed of the continental shelf' (The Truman Proclamation), 28 September 1945. See S.H. Lay *et al.* (compilers), *New Directions in the Law of the Sea* (Dobbs Ferry, N.Y.: Oceana Publications, Inc., 1973), Vol. I, pp. 103-6.

5. See 'Agreements between Chile, Ecuador and Peru', signed at the First Conference on the Exploitation and Conservation of the Maritime Resources of the South Pacific, Santiago (Chile), 18 August 1952, reproduced in Lay *et al.*, Vol. I, pp. 231-4.

6. See Luard, op. cit., pp. 30-42. See also Swing, op. cit.

7. 'Brunei: Southeast Asia's pocket size producer', *Petroleum News S.E.A.*, July 1973, pp. 38-42. See Proclamation dated 30 June 1954.

8. Position paper circulated by Philippine delegation among various states and the United Nations in 1955. See Arturo M. Tolentino, 'The waters around us: why the archipelagic doctrine is vital to the Philippines', BNFI Papers No. 5 (Manila, Philippines: Bureau of National and Foreign Information, undated), p. 12.

9. Luard, op. cit., p. 30.

10. See R.C. Amacher and R.J. Sweeney, eds., *The Law of the Sea: U.S. Interests and Alternatives* (Washington, D.C.: American Enterprise Institute for Public Policy Research, Inc., 1976), p. 62.

11. Luard, op. cit., p. 197.

12. Seyom and Brown, op. cit., p. 311.

13. Quoted in Amacher and Sweeney, op. cit., p. 92.

14. *Petroleum Economist*, June 1976, p. 235.

15. United Nations, Third Conference on the Law of the Sea, 'Revised single negotiating text', Part IV, Document A/CONF.62/WP.9/Rev. 2, 23 November 1976, reproduced in *Official Records*, Vol. VI, pp. 144-55.

16. Private source.

17. United Nations, Third Conference on the Law of the Sea, 'Informal composite negotiating text', Document A/CONF.62/WP.10/Corr. 2, 20 July 1977. All subsequent references to this will be 'the Composite Text'.

18. See Article 24, 'Convention on the territorial sea and the contiguous zone', Geneva, 1958, reproduced in U.S. Congress, Senate Foreign Relations Committee, *Legislation on Foreign Relations*, Joint Committee Print (Washington, D.C.: Government Printing Office, 1974), pp. 1239-46.

19. United Nations, Third Conference on the Law of the Sea, 'Revised single negotiating text', Document A/Conf. 62/WP.8/Rev. 1/Parts I, II, III (New York,

May 1976). All subsequent references to this will be 'the Revised Text'.

20. United Nations, Third Conference on the Law of the Sea, *Official Records* (New York, Caracas, and Geneva sessions), Vols. I, II, and III (New York, 1975). All subsequent references to this will be 'Official Records'.

21. *Petroleum News S.E.A.*, 31 May 1974, p. 5.

22. *Sunday Nation* (Singapore), 22 May 1977, p. 4, 'Viets make claim on off-shore zones', and U.S. State Department Communication, Singapore, 6 July 1977.

23. Proclamation, 1 September 1964, cited in U.S. State Department, Geographer, *Limits in the Seas* Series, No. 36. *National Claims to Maritime Jurisdiction*, third revision (Washington, D.C.: December 1975), hereinafter referred to as '*Limits in the Seas*'.

24. 'Declaration of 15 November 1968 by the Chairman of the Revolutionary Council of the Union of Burma on the Territorial Sea of the Union of Burma', in United Nations, Legislative Series, *National Legislation and Treaties Relating to the Territorial Sea, The Contiguous Zone, the Continental Shelf, the High Seas, and to Fishing and Conservation of the Living Resources of the Sea* (New York, 1970), p. 49.

25. *Asian Wall Street Journal* (Hong Kong), 12 April 1977, p. 6, 'Territorial waters of Burma extended' and 'Territorial sea and maritime zones law (Pyithu Hluttaw Law No. 3 of 1977)', promulgated on 9 April 1977, and reproduced in *The Working People's Daily* (Rangoon), 10 April 1977, pp. 1, 4.

26. 'Declaration du Gouvernement Royal en date du 27 Septembre 1969 relative a la mer territoriale et au plateau continental du Cambodge', in *U.N. Legislative Series*, op. cit., p. 51.

27. The term *geographically disadvantaged states* (GDS) has been used often in the literature on the Law of the Sea problem, but the exact definition has not yet been determined. The draft articles proposed by Haiti and Jamaica on the rights of developing geographically disadvantaged states within the economic zone or patrimonial sea offer a definition paraphrased as follows. The term refers to developing states which (a) are land-locked, or (b) for geographical, biological, or ecological reasons suffer from any or all of the following handicaps: (1) they derive no substantial economic advantage from establishing an economic zone or patrimonial sea, (2) they are adversely affected in their economies by the establishment of economic zones or patrimonial seas by other states, or (3) they have short coastlines and cannot extend uniformly their national jurisdiction. (Document A/Conf. 62/C.2/L.35, *Official Records*, Vol. III, p. 213.)

Additional examples of what may fall under (b) were offered the author; these are (1) states that cannot claim marine space to the extent allowed by the Convention, e.g., Singapore, or (2) states, although capable of extending their marine space, have no substantial resource. (Dr. Hasyim Djalal, Indonesian delegate to United Nations Law of the Sea Conference.)

28. Document A/Conf. 62/C.2/L.33, *Official Records*, Vol. III, pp. 212-13.

29. The term 'archipelago' is defined in the *Composite Text* as a 'group of islands, including parts of islands, interconnecting waters and other natural features which are so closely interrelated that such islands, waters and other natural features form an

intrinsic geographical, economic and political entity, or which historically have been regarded as such'. An archipelagic state would draw straight baselines joining the outermost points of the outermost islands and drying reefs, from which its territorial waters, contiguous zone, exclusive economic zone, and continental shelf would be measured. These baselines may not exceed certain limits, and waters within these baselines would be considered internal waters. See the *Composite Text*, Articles 46 to 54. See also *Official Records*, Vol. III, pp. 114, 226.

30. Cited in Peter Polomka, *ASEAN and the Law of the Sea* (Singapore: Institute of Southeast Asian Studies, 1975), p. 6. See also *Limits in the Seas*, p. 95.

31. See K. Romimohtarto *et al.*, op. cit., p. 76.

32. Information obtained during conversation on 3 June 1976, between the author and Dr. Hasyim Djalal, delegate to Law of the Sea Conference.

33. Idem.

34. *Petroleum News S.E.A.* (February 1977), p. 7.

35. See *Official Records*, Vol. II, pp. 23, 264-5; Vol. III, pp. 190, 202. See also Republic Act 3046 of 1961.

36. Tolentino, 'The waters around us', p. 3.

37. Arturo Tolentino, 'Archipelagic theory and the law of the sea', *The Philippine Geographical Journal*, Vol. XIX, No. 4 (Oct. – Nov. – Dec., 1975), p. 160.

38. Ibid., pp. 165-8. Also *Official Records* as cited.

39. See Document A/Conf. 62/C.2/L.49, *Official Records*, Vol. III, pp. 226-7.

40. See *Revised Text*, Part II, Article 14. See Article 46 in the *Composite Text*.

41. Tolentino, 'The waters around us', p. 4.

42. *Official Records*, Vol. II, p. 272.

43. For example, there is concern over Singapore's right to fish in what is at present considered 'high seas' but which would become Indonesian waters under the archipelagic principle. See *Official Records*, Vol. II, pp. 211, 268, and Doc. A/Conf. 62/C.2/L.33, *Official Records*, Vol. III, p. 212.

44. See discussion in Appendix B.

45. See Document A/Conf. 62/C.2/L.63, 15 August 1974, in *Official Records*, Vol. III, p. 233.

46. *Emergency (Essential Powers) Ordinance No. 7*, 2 August 1969, cited in *Limits in the Seas*, p. 122.

47. *Official Records*, Vol. II, p. 198.

48. See *Official Records*, Vol. II, pp. 198, 292-3.

49. See *Revised Text*, Part II, Article 119(7). Indonesia and Malaysia were reportedly not completely satisfied with the wording of Article 119(7) and jointly submitted a revised provision that would preserve rights traditionally exercised or existing under agreements covering archipelagic waters of a state, where these waters lie between parts of an immediately adjacent neighbouring state. (Private communi-

cation. Official documentation is not available at the time this is written.) See Article 47 (7) in the *Composite Text.*

50. Informal conversation between the author and the Indonesian delegate, 21 January 1976.

51. 'Convention on the continental shelf', United Nations Document A/Conf. 13/L.55, reproduced in U.S. Congress, Senate Foreign Relations Committee, *Legislation on Foreign Relations* (Washington, D.C.: Government Printing Office, 1974), pp. 1263-6. See Article 2.4.

52. Ibid., Article 1; italics added.

53. *Composite Text*, Article 76; italics added.

54. Ibid., Articles 77 and 81.

55. Ibid., Article 82.

56. Ibid., Articles 55 and 57.

57. Ibid., Articles 62 to 71.

58. Ibid., Article 121.

59. *Petroleum News S.E.A.*, April 1976, p. 21; J.D. Simmons, 'Developments in deep water drilling from floaters', paper presented at Offshore Southeast Asia Conference, SEAPEX, session, 18 February 1976; and *Asian Wall Street Journal*, 28 January 1977, p. 3, 'Exxon unit reports drilling at record depths off Thailand'.

60. Part of the boundary of the continental shelf area up to the 54th parallel was delimited by an agreement between the Netherlands and the Federal Republic of Germany on 1 December 1964. A similar agreement was signed between Denmark and the Federal Republic of Germany on 9 June 1965. See United Nations, *Proceedings of the Fourth Symposium on the Development of Petroleum Resources of Asia and the Far East* (New York, 1972), Vol. 1, p. 25.

61. Ibid.

62. The Court opined that the 'submarine areas concerned may be deemed to be actually part of the country over which the coastal state already has dominion — in the sense that, although covered with water, they are a prolongation or continuation of that territory, an extension of it under the sea'. (International Court of Justice, North Sea Continental Shelf Cases, Judgement of 20 February 1969, paragraph 46.) See Lay *et al.*, op. cit., Vol. I, p. 193.

63. U.S. Department of State, Geographer, *Limits in the Seas*, No. 46, *Theoretical Area Allocation of Seabed to Coastal States*, International Boundary Study, Series A, 12 August 1972. Cited in Amacher and Sweeney, op. cit., p. 30.

64. United Nations, General Assembly, *Report of the Committee on the Peaceful Uses of the Sea-Bed and the Ocean Floor Beyond the Limits of National Jurisdiction*, Official Record, 27th Session, Supplement No. 21 (New York, 1972), p. 45.

65. See U.N. Document A/C.1/L.632/Rev. 1, dated 6 December 1972.

66. See U.N. Document A/AC.138/87, dated 4 June 1973, 'Economic significance, in terms of sea-bed mineral resources of the various limits proposed for national juris-

diction', prepared by the Committee on the Peaceful Uses of the Sea-Bed and the Ocean Floor beyond the Limits of National Jurisdiction.

67. Luard, op. cit., p. 257.

68. Actual conflicts over territorial claims in the Gulf of Thailand are discussed in the next chapter.

69. 'Proclamation concerning the Indonesian Shelf', 1 February 1969, submitted as Document CCOP (VI)/4, and published in ECAFE, CCOP, *Report of the Sixth Session*, 1969, p. 136.

70. Conversation between the author and an Indonesian Delegate to the Law of the Sea Conference, Dr. Hasyim Djalal, 3 June 1976. See also *New Directions in the Law of the Sea*, Vol. IV, pp. 91-104, for the Australia-Indonesia agreements, and *Limits in the Seas*, p. 95, for the agreement with India.

71. *Official Records*, Vol. II, p. 169.

72. Ibid., pp. 198, 278.

73. Published in United Nations, *Legislative Series*, op. cit., pp. 375-9.

74. Malaysia has ratified a tripartite agreement with Indonesia and Thailand on the Straits of Malacca shelf (see above). It has discussed shelf boundaries on the Gulf with Thailand but by late 1976 no agreement had been ratified. See *Petroleum News S.E.A.*, August 1976.

75. *Official Records*, Vol. II, pp. 155, 224.

76. *New Straits Times* (Kuala Lumpur), 12 March 1975.

77. *Straits Times* (Singapore), 11 October 1976, p. 3.

78. *Asian Wall Street Journal* (Hong Kong), 12 April 1977, p. 6, and 'Territorial sea and maritime zones law' in *The Working People's Daily* (Rangoon), 10 April 1977, pp. 1, 4.

79. See Chapter 4 of Law No. 3 of 1977.

80. 'Declaration du Gouvernement Royal en date du 27 Septembre 1969 relative a la mer territoriale et au plateau continental du Cambodge', reproduced in United Nations, *Legislative Series*, op. cit., p. 51.

81. See discussion on semi-closed seas in Appendix B.

82. *Official Records*, Vol. II, pp. 22, 192-3, 225.

83. Ibid., Vol. II, p. 159.

84. 'Proclamation from the President dated 7 September 1967 about the policy of the Republic of Viet-Nam concerning the sub-soil, seabed and resources of the Continental Shelf', submitted as Document I & N R/R.98 and published in ECAFE, CCOP, *Report of the Fifth Session*, 1968, p. 151.

85. *Official Records*, Vol. II, p. 162.

86. *Sunday Nation* (Singapore), 22 May 1977, op. cit.

87. U.S. State Department communication, Singapore, 6 July 1977.

88. *Sunday Nation*, 22 May 1977, op. cit.

89. Ibid. Italics added.

90. Proclamation No. 370 of 20 March 1968 by the President of the Philippines.

91. See *Far Eastern Economic Review*, 28 May 1976, p. 115; *Straits Times* (Singapore), 18 June 1976, p. 2, *Straits Times* (Singapore), 21 June 1976, p. 4; *Philippine Daily Express* (Manila), 31 July 1976, p. 18; and *Straits Times* (Singapore), 6 July 1976, p. 8. The Philippines has also used the *res nullius* and national security arguments.

92. See *Straits Times*, 15 March 1976.

93. *Official Records*, Vol. I, pp. 134, 135 and Vol. II, pp. 151, 211, 285.

94. *Asian Wall Street Journal* (Hong Kong), 'Sea law ends as members divide on progress', 20 September 1976, pp. 1, 7. See also *Official Records*, Vol. VI, pp. 133-5.

95. *Sunday Times* (Singapore), 'Sea law talks end without pact', 17 July 1977, p. 2.

IV

Actual Territorial Disputes over Potential Off-shore Petroleum Fields

PROPERTY rights over islands or shelf areas in the Gulf of Thailand and the South China Sea have become sources of open conflict or rancour among the coastal countries. These claims may or may not necessarily be settled by resolution of the issues in the Law of the Sea Conference, as ownership claims are partially based on previously acquired rights. The discussion here will be grouped under three headings: (1) boundary disputes in the Gulf of Thailand, (2) controversial claims to sovereignty over islands in the South China Sea, (3) the Philippine claim to Sabah. Although not originally linked to petroleum, the Sabah question may now be relevant in the light of off-shore discoveries.

Controversial Claims in the Gulf of Thailand

From the early 1970s up to the fall of the previous South Vietnamese and Cambodian governments in early 1975, the Gulf of Thailand was the scene of conflicting national claims over areas of the continental shelf leased out for petroleum exploration/exploitation by private companies. The disputes were (1) between Cambodia and Thailand in the northern portion where Cambodia and Thailand share the shelf; (2) between Cambodia and Vietnam where these two countries share the shelf in the southern part; and (3) among all three countries in the central part of the Gulf where the three-country claims to the shelf converge (see Fig. 5). Including the tripartite overlap, the Thai-Cambodian dispute involved about 15 600 square kilometres.[1] The strictly Thai-Cambodian dispute involved about 8 000 square kilometres.[2] The controversy between Cambodia and South Vietnam — the most extensive area of dispute in the Gulf — covered about 45 000 square kilometres, with about two-thirds of the Cambodian concessions overlapping South Vietnamese concession blocks.[3] Ideally, countries should put out offers to bid on concessions only when boundaries are clear and not subject to controversies or claims. This has not been the case, however, because of the eagerness of each country to get exploration started and underway.

Although Thailand's Royal Decree concerning its shelf was not approved until mid-1973, announcements of the bases for awarding the right to explore for and produce oil in the Gulf of Thailand were issued by the Thai government on 21 June 1967, at which time bids were invited.[4] This invitation followed the drafting of the Thai Petroleum Act and Petroleum Income Tax in 1967[5] which received legislative approval only in April 1971.[6] Awards in the Gulf of Thailand and Andaman Sea were made in 1968,[7] straddling areas claimed by Cambodia and South Vietnam.

In November 1969 the government of Cambodia reached an agreement with Elf Acquitaine for exploration rights on the off-shore shelf claimed by the country.[8] On 10 August 1972, the Cambodian Ministry of Industry and Mining Resources officially invited new bids for both on-shore and off-shore areas.[9] In July 1973 the Ministry awarded an off-shore concession (part of an area relinquished by the Elf/Exxon[10] consortium) to Marine Associates, the southern half of which overlapped the shelf claimed by South Vietnam.[11]

THE BASES OF INDIVIDUAL COUNTRY CLAIMS

The controversies involved, primarily, historical claims to sovereignty over small islands and, to some extent, interpretation of the continental shelf concept in delimiting the seaward distance of shelf boundaries, notwithstanding observations to the contrary. Between Thailand and Cambodia, the dispute was over the island of Koh Kut. Between South Vietnam and Cambodia, the contention was on sovereignty over the islands of Pulau Wai, Pulau Panjang, and the larger Phu Quoc.[12]

Cambodia based its off-shore claims on the French-Siamese Treaty of 23 March 1907, a subsequent verbal understanding of 8 February 1909 (that delimited the Siamese-Cambodian border), and application of clauses of the Geneva Convention on the Continental

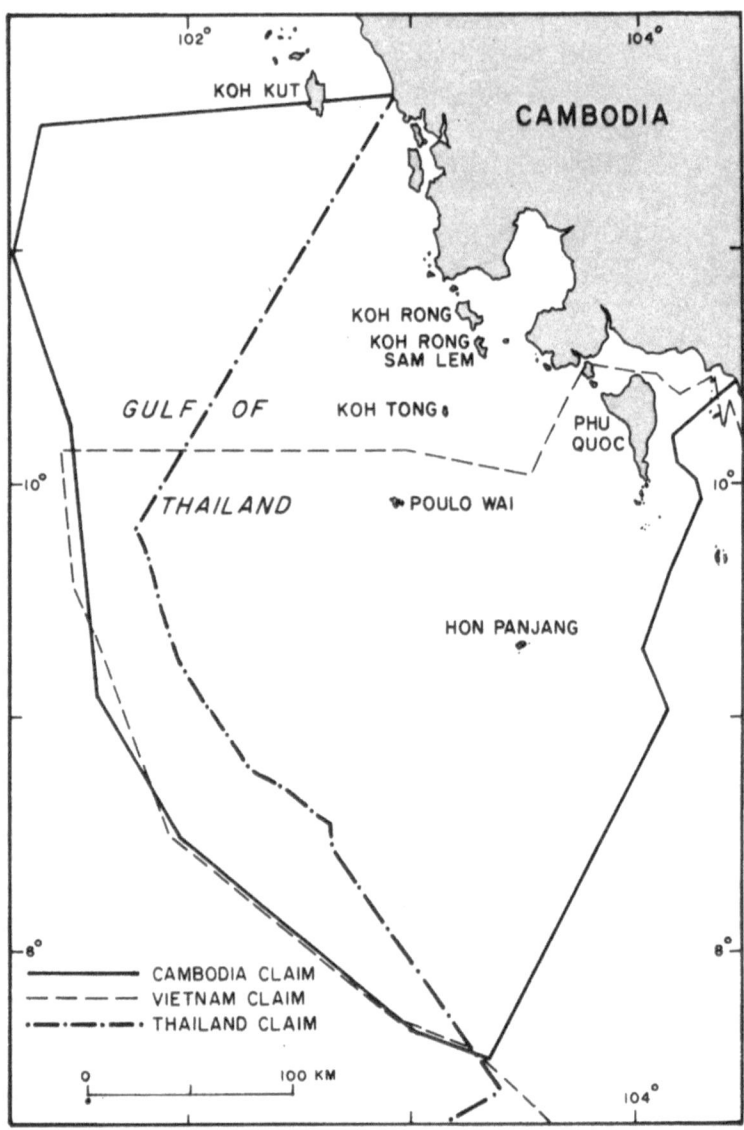

Fig. 5. Tripartite Gulf of Thailand Shelf Dispute
(Modified version of map in *Petroleum News S.E.A.*, June 1973;
used with permission.)

Shelf.[13] The Cambodian government declared the state's full and absolute sovereignty over its continental shelf on 27 September 1969.[14] On 3 July 1971, Cambodia announced the coordinates of its continental shelf. These coordinates were given as 102-deg., 54-min. E. Long., and 11-deg., 38-min., N. Lat., and excluded the island of Koh Kut.[15] On 1 July 1972, however, Presidential Decree or *Kret* No. 439-72/PRK included part of Koh Kut island within the Cambodian territorial area as well as some shelf area associated with Koh Kut, using the summit of this island as the reference for territorial division of the island.[16] Vietnam is reported to have advocated the mid-channel between Pulau Wai and Koh Tang as the boundary between Cambodia and Thailand.[17]

The South Vietnamese government's claims on the islands are supposed to be based on historical occupancy.[18] Thailand's territorial claims over Koh Kut are apparently based on application of the Convention, as indicated by the Royal Decree delimiting its shelf area and extending exploration rights in the Gulf of Thailand.[19] It is also easy to see from the overlap of claims in the central part of the Gulf that part of the dispute between the three countries arises from independent applications of the continental shelf concept for measuring shelf boundaries in areas which are adjacent states or boundaries of states whose coasts are opposite each other. The Convention provides that in the case of two adjacent states, with the *same* continental shelf, the boundary would be determined by agreement between them or by application of the 'principle of equidistance from the nearest points of the baselines from which the breadth of the territorial sea of each State is measured' (Article 6.2). For two or more states whose coasts are opposite each other and sharing the same continental shelf, the boundary would be determined by agreement or 'the median line, every point of which is equidistant from the nearest points of the baselines from which the breadth of the territorial sea of each State is measured' unless 'another boundary line is justified by special circumstances' (Article 6.1). The 'special circumstances' could be previously acquired or historical rights. The 'median line' concept for boundary delimitation in the absence of agreement appears to have been ignored (see Fig. 6). Of course, in principle, the provision on boundaries in the Geneva Convention

80

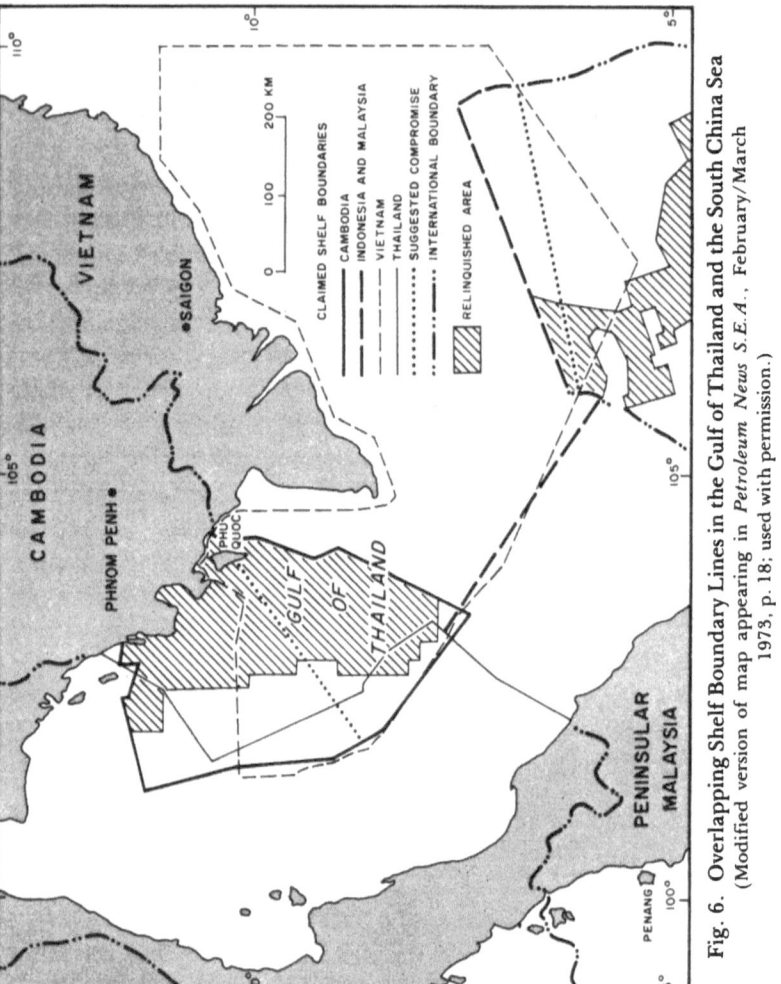

Fig. 6. Overlapping Shelf Boundary Lines in the Gulf of Thailand and the South China Sea (Modified version of map appearing in *Petroleum News S.E.A.*, February/March 1973, p. 18; used with permission.)

on the Continental Shelf is not necessarily binding on a state that was not a party to the Convention. As stated earlier, in South-East Asia, only Cambodia, Malaysia, and Thailand ratified the Convention. South Vietnam did not.[20]

STALEMATE IN NEGOTIATIONS

Reports of the progress of negotiations and points of agreement appeared in the press in subsequent years. There were no really clear indications, however, that the problem was close to solution.

In October 1972 it was reported that there were indications that South Vietnam would soon actively seek a boundary agreement with Cambodia prior to seeking bids on off-shore acreage. Thailand was reported to be similarly ready to seek an agreement with Cambodia on their mutual boundary. However, on 1 July 1972, Cambodia extended the area of its claim northward, citing historical evidence.[21]

In mid-1973 the Tenneco group (Tenneco-Marathon-Phillips-Agip), that had a promising oil/gas strike in a wildcat well in 240 feet (73 metres) of water 135 miles (217 kilometres) from the shore of Thailand, announced it would drill four more wells to confirm the size of its discovery. Test rates on the first well were, however, withheld pending the conclusion of boundary negotiations between Thailand, Cambodia, and South Vietnam to solve overlapping acreage claims near the discovery.[22] The initial oil and gas shows were reported to be located within 25 kilometres of the disputed border claims.[23]

By February 1974, observers noted that 'spasmodic working-level negotiations' had been taking place over the years between the governments of Cambodia, Thailand, and South Vietnam but with no positive results.[24]

In the case of the Cambodian-South Vietnamese dispute, conflicting reports appeared over several months on what each country was willing to keep or relinquish. Observers noted in 1973 that the most important of the islands disputed was Phu Quoc, largely occupied by South Vietnamese but long claimed by Cambodia. It was thought that the island might be excluded from concession area offers in order to facilitate a compromise on the remaining disputed area. There were also reports in early 1973, supposedly reliable, that a common boundary between Cambodia and South Vietnam was 'more or less agreed upon'.[25]

In February 1974 it was reported that the South Vietnamese Petroleum Board had made a tentative proposal to the Cambodian Directorate of Mines which would roughly bisect the 45 000 square kilometre overlap. The South Vietnamese government would give up its claim to a triangular area of about 24 000 square kilometres on the north-west. This would give Pulau Wai to Cambodia, while Vietnam would retain Pulau Panjang and Phu Quoc. It was reported at that time that the Cambodian government had 'informally rejected' the proposed compromise.[26] In June 1974 it was reported that the tripartite boundary dispute in the Gulf of Thailand was at a standstill as the three countries reached an agreement not to drill in the disputed zone until the conflict had been resolved.[27]

In September 1974, however, drilling off Pulau Wai by an oil consortium in the area of the Cambodian-Vietnamese dispute, on a concession awarded by the Cambodian government, drew a vigorous protest and a threat of violence by the South Vietnamese government. A comment on the incident explained that the Saigon government's ultimatum 'was in line with claims by Saigon that the marine boundary between the two nations should be drawn so as to leave Pulau Wai in South Vietnamese waters'.[28] This incident, and the explanation, appeared to contradict reports in early 1974 on the fate of Pulau Wai.

The Saigon government was reported to have threatened to dismantle the *Glomar IV* drilling rig if it did not voluntarily leave the Wai island area in the Gulf of Thailand. Press reports stated the South Vietnamese government delivered a note to the Cambodian embassy in Saigon early in September, advising that the Vietnamese navy was prepared to dismantle the rig if it were not removed by 12 September. The Cambodian government responded by sending navy boats to patrol the region and sent a battalion of 300 marines to reinforce a navy base in the area. On 6 September 1974, the South Vietnamese government cabled the Cambodian capital to express its wish to settle the dispute through peaceful negotiations. Nevertheless, Elf/Erap ordered the *Glomar IV* to stop drilling before the 12 September deadline in the interest of the crew's safety. Negotiations

were reported to have taken place between the two governments.[29]
` Later reports indicated that Vietnamese troops had thrown the
Cambodians off Pulau Wai shortly after the fall of Saigon. Official
sources in Vietnam were, however, reported to have informed in-
terviewers in October 1975 that the islands were occupied by Cambodians. The government of unified Vietnam reportedly would keep
its claim but preferred to settle all such disputes amicably.[30]

Oil-related Area Claims in the South China Sea

Controversial oil-related claims to shelf and island areas in the
South China Sea involve at least five parties — Vietnam, China, Taiwan, the Philippines, and Indonesia.[31] Vietnam has at least three oil-
related disputes, two of which have included armed or strident verbal
conflicts. Currently the most serious source of conflict affecting the
development of South-East Asia's oil resources relates to sovereignty
disputes over the Spratly Archipelago. For this reason, more space
will be devoted to discussion of the background and current developments in this area than to other South China Sea conflicts. Since
some of the sovereignty disagreements in the southernmost part of
the South China Sea are related to claims over the Spratlys, a discussion of such disagreements will be included at the end of this
section.

THE SPRATLY ARCHIPELAGO AND THE BACKGROUND TO CLAIMS[32]

The Spratly Archipelago, as this group of islands is now described,
is made up of more than 100 islands, keys, atolls, and shallow banks,
scattered over an area of 180 000 square kilometres. Geologically
these islands and islets are separated from the shelf areas of the South
China Sea's littoral states by troughs and abyssal plains.[33] There are
four countries directly claiming ownership of one or all of these
islands: China, Taiwan, Vietnam, and the Philippines. The first
three claim ownership of all the islands on the (historical) basis of
prior discovery and occupation, directly or indirectly. The Philippines claims ownership to only a few islands, 'discovered' and occupied only after the Second World War. The Philippines occupies

five, Vietnam six and Taiwan one of the islands; China occupies none.

The Spratly Archipelago has grown in importance since the government of South Vietnam awarded eight blocks of oil concessions in July 1973 to four foreign consortia in the waters south and southeast of Phuoc Tuy province. The reasons are: its natural resources, especially its potential petroleum resources; and its strategic location (a) between the Indian Ocean and the West Pacific, and (b) relative to its proximity to the Philippines.

Spratly Island alone is located in the south-west part of the archipelago and is about 550 kilometres due west of the westernmost tip of Palawan island in the Philippines Archipelago (see Fig. 7). It is about 650 kilometres south-east of the coast of Vung Tau in Vietnam, about 950 kilometres south of the island of Hainan, off mainland China, and about 1 450 kilometres south-west of the southern tip of Taiwan. Approximately 800 kilometres north of Spratly Island are the Paracels.

For centuries the South China Sea islands were largely uninhabited although visited by fishermen. They were sanctuaries for bird-life and sea turtles and were of economic interest primarily because of guano, phosphate resources, and fish-life. The Spratlys were named after the captain of a British whaling vessel that landed there in 1840. The Japanese were reported to have actively undertaken guano exploitation in Pratas, until the Chinese government bought the islands back and installed territorial markers in 1909. The Japanese then engaged in guano mining operations in both the Paracels and the Spratlys, until they were reportedly dislodged from the Paracels by Chinese warships and from the Spratlys by French warships.[34] The Japanese were again said to have operated a phosphate mine in the Spratlys during the Second World War, and also used both the Spratlys and the Paracels as garrisons, refueling stations and submarine bases.[35]

The Chinese claim.[36] The governments in Peking and Taipei have consistently laid claim to four groups of islands in the South China Sea: (1) the Spratlys which they call Nansha, (2) the Paracels which they call Hsisha, (3) the Macclesfield bank which they call Chungsha,

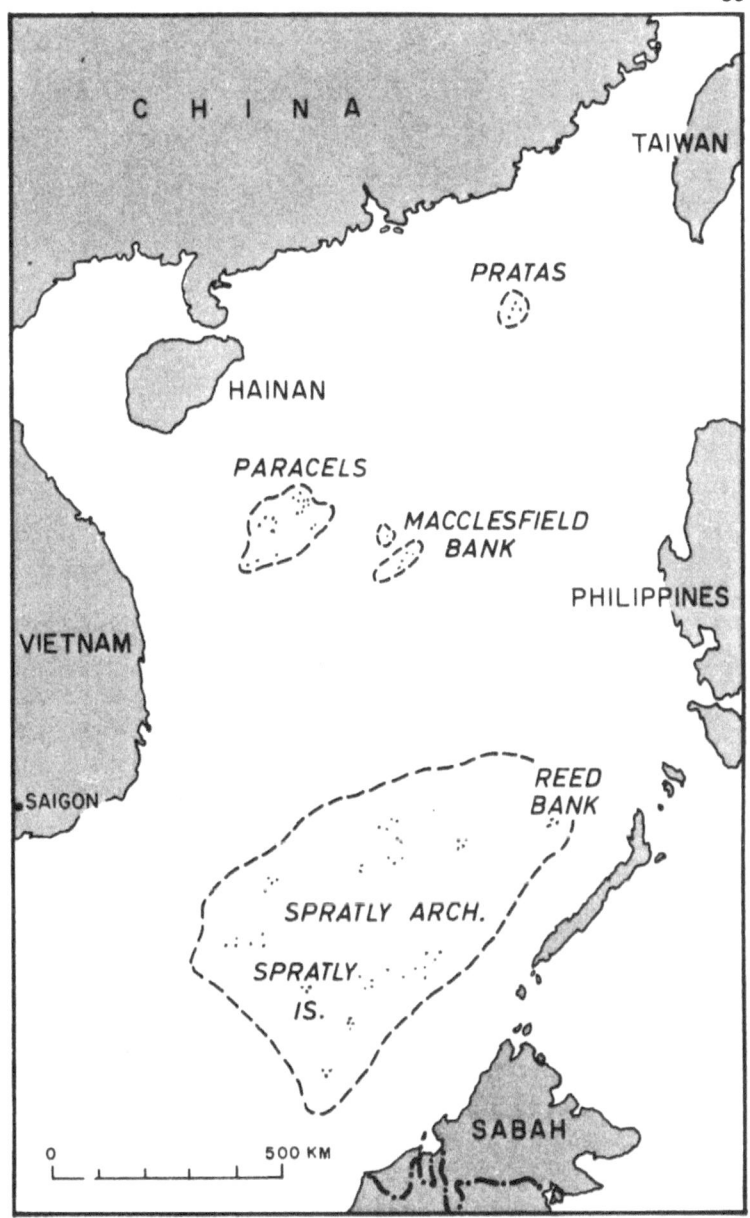

Fig. 7. Spratly and Paracel Archipelagos

and (4) Pratas Island which they refer to as Tungsha. (Translated the Chinese names mean, respectively, Southern Sands, Western Sands, Middle Sands, and Eastern Sands. See Fig. 7 again.)

The Peking claim is reportedly based on rights that date back to as early as 200 B.C.,[37] and is said to be supported by many historical records, including maps, and evidence of early Chinese settlements. One point of reference is the presence of Chinese merchants in places further south, such as Borneo and Java, as early as the first century B.C.[38] Another point of reference is the presence of Chinese vessels in the area dating back to the Han period (206 B.C.–200 A.D.). The islands had also been included by the Chinese as part of their territory from at least the Sung period (960 – 1279 A.D.), with some accounts indicating an earlier period. Household wares and other relics dating from the Tang Dynasty and Northern Sung period were reported to have been uncovered in the Paracels. The Taipei government dates its claim to the period 1404–33 when Cheng Ho made voyages to South-East Asia.[39]

Post-Second World War claims by the Chinese to the Spratlys and Paracels were made as early as 1951.[40] Although the San Francisco Peace Treaty divested Japan of its 'rights, title and claim to the Spratly Islands and the Paracel Islands',[41] it did not vest the same elsewhere. Neither the then new Communist government (People's Republic of China) in Peking nor the Nationalist government in Taiwan participated in the Conference, and the government in Peking denounced the peace conference and treaty as 'illegal', while reiterating Chinese ownership of the two groups of islands. During the Conference, the Vietnamese delegation issued a statement 'confirming' the two archipelagos as Vietnamese territory. Taiwan also claimed ownership by right of occupation under international law.

On 1 June 1956, shortly after a Philippine national staked his claim on some islands in the area, the *Kuangming Daily* in Peking declared that the Chinese had discovered the Nansha Islands even before Magellan 'discovered' the Philippines. Although the Nationalist government in Taipei had originally withdrawn its troops from all the South China Sea islands in 1949, it is known that by 1956 it had troops occupying the Itu Aba island in the Spratly Archipelago.[42] It also had set up a weather station there.

On 11 January 1974, a few months after Saigon had awarded oil exploration concessions in the South China Sea and then announced the incorporation of more than ten islands in the Spratly Archipelago within its territory, China issued another strong statement reiterating its claim to the South China Sea islands and their natural resources.[43] China's claim to the Paracels and Spratlys was reasserted in October and November 1975. An article in the *Kuangming* official daily on 24 November 1975 presented historical documents and archaeological evidence to support its claims. A related film based on this article indicated Chinese presence in the four island groups and the work of a PRC archaeological team in the Paracels which had uncovered Chinese coins, porcelain, and other artifacts which purportedly dated from the Tang (618–907 A.D.) and Sung (960–1279 A.D.) periods.

Aside from the question of economic resources, China's claims to the South China Sea Islands may be viewed in relation to the Sino-Soviet dispute.[44] The Spratlys are thought to be strategically important because of their proximity to the Straits of Malacca; their control would be essential to the protection of the sea routes of China's sizeable and growing merchant fleet. In addition, with the Soviet naval build-up in the Indian Ocean and the Pacific, the control of the Paracels and Spratlys would provide potential locations on China's south flank for advance monitoring of ship movements and bases for reconnaissance craft.

Vietnam's claim. The former South Vietnam government claimed that the Spratly Islands officially became part of Indochina in 1834 when the Emperor Minh Mang recorded the islands as part of Vietnamese territory. The French occupied the Paracel and Spratly islands between 1922 and 1933, and claimed ownership under international law. It claimed that occupation under the law required both administrative and military control of the state, both of which it had fulfilled.[45] Shortly after Filipino Tomas Cloma's announcement of his claim to some islands, the government of South Vietnam issued a communique on 1 June 1956, stating that the Spratly Islands, along with the Paracels, had 'always been a part of Vietnam', and that this was recognized in the San Francisco Peace Conference in September 1951.

On 9 June 1956 the French Charge d'Affaires in Manila told the Philippine Department of Foreign Affairs that by virtue of occupation between 1923 and 1933 the Spratlys belonged to France and that, unlike the Paracels, they had not been ceded to Vietnam. This claim was disputed by a spokesman of the South Vietnam legation, who stated that the Spratlys were officially incorporated into the Vietnamese province of Baria in 1929, and that at the San Francisco Peace Conference nobody had contested the declaration by the Vietnamese delegation that Vietnam had sovereignty over both the Spratlys and the Paracels.

The Saigon government incorporated the Spratlys as part of the province of Phuoc Tuy by decree in September 1973, shortly after awarding exploration concessions in that region to several oil consortia. Exploration operations were cut short by the fall of the incumbent South Vietnam government in April 1975.

On 6 May 1975 the Hanoi News Agency reported that the new South Vietnamese government had occupied six islands in the Spratly Archipelago. A subsequent territorial map of a reunified Vietnam published in late 1975 also included the Paracels and the Spratly Archipelago (known in Vietnamese as Quan Dao Hoang Sa and Quan Dao Truong Sa, respectively).[46] On 5 June 1976, the Provisional Revolutionary Government of South Vietnam stated 'once again its sovereignty with regard to the Truong Sa (Spratly) Islands and [reserved] for itself the right to protect that sovereignty'. On 26 July 1976, it was reported that the first postage stamp issued in the country since reunification of the North and the South showed a map of the new Vietnam that included the Paracel and Spratly Islands.

The Philippine claim and clashes with other countries. Philippine involvement in the Spratly group of islands dates back to the period 1947–50, when Tomas Cloma, a Philippine owner of a fishing fleet 'discovered' a group of islands and islets in an area some 650 kilometres west of Palawan island in the Philippines. On 15 May 1956, Cloma's group hoisted the Philippine flag on one of the unoccupied islands and formally staked a claim. Cloma called the area of about 170 000 square kilometres 'Freedomland'.

The publicity accompanying the claim drew hostile reactions from China, Saigon, and Taiwan. Taiwan announced it was 'possible and probable' that it would dispatch troops to the Spratlys. In subsequent expeditions, Cloma's men found that markers they had put up on Itu Aba Island had been removed and replaced by Chinese signs. In October 1956, Taiwan government troops clashed with Cloma's group, after which Cloma's Freedomland excluded the Spratly Island 'in deference to Nationalist China'.

Over the past twenty years, the Philippine government has used at least four arguments to justify its occupation of Freedomland and, recently, exploration of the Reed Bank: (1) the *res nullius*[47] principle; (2) rights acquired under the Japanese Peace Treaty in San Francisco in 1951; (3) the national security of the Philippines; and (4) more recently, in connection with drilling in the Reed Bank, rights acquired under the 1958 Geneva Convention on the Continental Shelf.[48]

At the time Cloma staked his claim in 1956, the Philippine Department of Foreign Affairs stated that the Philippine government regarded the islands, islets, etc. within Freedomland as *res nullius*, that some of them were 'newly-risen'; therefore, they were available for economic exploration and settlement by Philippine nationals under international law. It was also argued that the Spratlys (and the Paracels) had been turned over to the Allied Powers by Japan in the Peace Treaty signed in San Francisco on 8 September 1951, but disposition of the territories involved had remained unsettled.

At the Constitutional Convention in 1971, a Batanes province delegate sponsored the Philippine claim on Freedomland and the Spratly islands, basing his sponsorship on the issue of national security of the Philippines and also asserting the modes of acquiring a territory under international law through discovery, occupation and possession. On 10 July 1971, the Philippine government issued a formal press statement to the effect that occupation of the island of Itu Aba by Taiwan's forces constituted a threat to the national security of the Philippines. The island, it was stated, was located '480 miles from Manila, 960 miles from Taipei and 470 miles from Saigon', and had been used during the Second World War as a staging area in the invasion of the Philippines by the Japanese. The Philippine government also requested Taipei to withdraw its soldiers from Itu Aba, and stationed its own soldiers in three other islands in the area. A few days following this statement, the governments of Great Britain and the Netherlands announced the abandonment of their rights as trustees over the Spratlys. The Philippine claim was reiterated at the 72nd meeting of the Sea-Bed Committee of the United Nations in March 1972.

In early 1974 the Philippine government sent a diplomatic protest to the governments of South Vietnam and Taiwan. It protested against the display of force used by both parties in laying claim to the Spratlys and nearby islands, and urged that the question involving ownership of the contested territories be addressed to the United Nations or to the Allies of the Second World War for resolution.

In that note the Philippine government claimed (1) that the five islands occupied by its soldiers had been acquired by right of occupation; (2) that the islands did not form part of the Spratly Island Group, as they were located about '200 miles to the north-east' of the Spratlys and that Puerto Princesa (capital of the province of Palawan) lies directly east by about '250 miles';[49] (3) that the sovereignty over the islands, which were *res nullius*, had been acquired by occupancy; and (4) that its location rendered it strategically important to Philippine national security.

OIL DRILLING IN THE REED BANK

The strident reaction of China and Vietnam to the drilling by a Swedish-Philippine consortium in the South China Sea in 1976 made the Spratly Archipelago a potential political volcano. Early in 1976 a group of Swedish and Philippine companies signed a contract with the Petroleum Board of the Philippines (now the Energy Development Board) to explore on 15 000 square kilometres on the Reed Bank, part of the Spratly chain and some 260 kilometres due west of the Philippine territorial baseline. The first known reaction from one of the contending claimants to the Spratlys came from Vietnam, about a week after the first wildcat well (Sampaguita No. 1) was drilled. On 5 June 1976, Vietnam reiterated its 'sovereignty' with regard to the Spratlys and its right to protect that sovereignty.

On 16 June 1976, Peking issued a strongly worded statement repeating its claim to sovereignty over the Spratlys. Its declaration included the following:

It was recently announced by official Philippine sources that a consortium of Swedish-Filipino oil exploration firms had started oil drilling operations in the area of Liletan (Reed Bank) of China's Nansha Islands ....

The Nansha Islands ... [have] always been part of China's territory.

Any foreign country's armed invasion and occupation of any of the Nansha Islands or exploration and exploitation of oil and other resources in the Nansha Island area constitute encroachments on China's territorial integrity and sovereignty and are impermissible.[50]

The statement was directly aimed at the Philippines. At that point the Sampaguita No. 1 well had reached a depth of over 3 000 metres.[51]

The Philippine government responded by stating that the exploration activity did not encroach on any foreign country's territory as it considered the Reed Bank to be within the continental shelf of the Philippines and that its economic exploitation was conducted within the provisions of the 1958 Geneva Convention on the Continental Shelf. (Actually, although two-thirds of the contract area is at depths of less than 100 metres, the present drilling is west of the Palawan Passage where the reported deep trench breaks the continental shelf west of Palawan. See discussion in Chapter III.)

A columnist of the *Daily Express* in Manila (said to reflect official sentiment) also restated the Philippine government's position that it was claiming jurisdiction only over 'the islands they actively occupy'.[52] Nevertheless, the Philippine Foreign Secretary announced, on 20 June 1976, that the subject of ownership of the Spratly Islands would be discussed during his visit to Peking in August 1976. A month after, the Philippine government announced that the visit had been delayed and a new date would be arranged later.[53]

The drilling of the Reed Bank well had, in fact, become a touchy subject as early as May 1976; at that time the *Far Eastern Economic Review* reported that the top officials of the Philippine National Oil Company refused to divulge the exact location of the well, on 'orders of Malacanang'.[54] With oil-bearing zones reported to have been found in the first well,[55] the controversy over ownership over this area is unlikely to die out.

Some press articles report that the Philippine government is bank-
ing on the fact that possession is 90 per cent of the law.[56] This argu
ment holds if one can protect one's right of occupancy against someone else who should want ownership of that territory. This implies
that one has the military strength to protect that right. For example,
should China insist, by military force, that its historical claims be
upheld, this could ignite the 'fuse that could set off an international
explosion'.[57]

In the meantime, Taiwan was reported to have reiterated officially
in late 1976 that the Spratlys group was part of Taiwan's territory,
and that any decision reached by any foreign country related to it
was 'null and void'.[58]

Suggestions for dealing with the four-country claim to the Spratlys
by dividing the South China Sea among the coastal countries, in
the manner in which the North Sea has been divided among the
European nations bordering that body of water,[59] appear to ignore
the facts of the North Sea case. One of the main features of the ruling
of the International Court of Justice was that it advocated the application of the natural prolongation principle to the delimitation of
the boundaries of the Continental Shelf.[60] In addition, the equidistance principle is applicable only where the same continental
shelf is an extension of a land mass occupied by several coastal states,
so that the shelf falls under more than one territorial jurisdiction.[61]
In any other instance, under present laws, the seabed beyond territorial waters and beyond the continental margin of the coastal state
would be open to all nations, unless it is already owned by somebody
else.

There is, of course, the question of the validity of China's claims
to all the islands. Questioning this validity does not solve the problem, however, if China wants to pursue its claims heatedly and
with military force. Should one question the historical basis for its
claims, one could equally contest the historical basis for any nation's
claims, including the Philippine claim to historical waters.

The only pragmatic answer would appear to be the resolution
of the problem at a treaty table, and to abide by whatever agreement is reached. Such an agreement is conceivable. A North Vietnamese official was quoted in 1974 as indicating 'a compelling need

for negotiations' as a result of discovery of off-shore oil in the region.[62] It is, however, beyond the scope of this study or the intention here to delve into resolution of this problem. These are questions for political scientists and international jurists.

CONTROVERSY OVER OTHER PARTS OF THE CHINA SEA

Two other relatively minor trouble spots exist in the South China Sea. One is the Paracel Islands, claimed by both Vietnam and Peking. The other, in the south-west portion of the Sea, is between Indonesia and China, on the one hand, and between Indonesia and Vietnam, on the other. A potential problem also exists between China and Malaysia, as the former has extended its South China Sea claims as far as the Tsengmu Reef on the shelf of Sarawak.[63]

The Vietnamese claim to the Paracels goes back to 1802 when Emperor Gia Long established a company to control and exploit that archipelago's natural resources.[64] Vietnamese-styled pagodas have reportedly been found in the Paracels.[65] The Vietnamese also claim that the French occupied the Paracels in the early 1930s 'in the name of the Vietnamese Royal Kingdom' and transferred them as an administrative unit to the Vietnamese kingdom shortly thereafter.[66] French occupation lasted until the eve of the Second World War, and Vietnamese claims are in part based on an extension of the French claim to the islands. South Vietnam announced the incorporation of the Paracels into Quang Nam Province in 1961, and announced the presence of a meteorological tower on at least one of the Paracels.[67] It also announced that it had regularly stationed troops, patrolled the sea areas, and exercised administrative control over the area.

The claims of China to the Paracels were discussed earlier in connection with its claims over the Spratlys. After the Second World War, China also set up an observation and communications post on Woody Island in the Paracel group, stationed troops in other Paracel islands, and patrolled the neighbouring seas.[68]

In 1973, shortly after the former South Vietnam government issued exploration permits in the Spratlys, it set up a garrison in the Paracels. On 11 January 1974, the Foreign Ministry of China issued a statement reaffirming its 'indisputable sovereignty' over all four

groups of islands in the South China Sea, and over 'the natural resources in the seas around them'.[69] Saigon refuted this claim the
next day, and both countries attempted to enforce their claims.
China's soldiers planted Chinese flags on Robert Island, an act Saigon condemned on 16 January. Saigon also sent troops to take down
the Chinese flags on 17 January, with no incident. At the same time
Saigon presented a complaint against the Chinese action to the
United Nations Security Council. There were skirmishes at sea on
18 January, followed by a land confrontation on Duncan Island on
the 19th. Air and land battles continued on two other Paracel
islands— Pattle and Money— until China's troops gained full control
of the islands.[70]

The seizure of the Paracels by China's troops was reported to have
incensed the governments not only of South Vietnam, the Philippines, and Taiwan but of Indonesia as well. Indonesia was reported to
have objected to what it considered were China's attempts to extend its
possessions throughout the South Pacific Basin, all the way down to a
small island group in Indonesian territory ending just north of Sarawak. Indonesia cited as indications of China's intentions a map
and accompanying text in a 1971 issue of *China Reconstructs*.[71]
In September 1976, China was reported to be drilling for oil in the
Paracels which had been converted into a 'steel fortress'.[72]

A boundary dispute also exists between Indonesia and South
Vietnam in the South China Sea. In 1968 Pertamina awarded an
off-shore concession block to a group of three companies (Agip,
Tenneco and Phillips) in the northernmost part of what it regarded
as the Indonesian shelf. However, the shelf area claimed by South
Vietnam and opened for bidding in 1973 overlapped this block. In
early 1973 it was reported that the South Vietnam government was
amenable to an approximate east-west boundary line that would
somewhat halve the disputed area.[73] Nevertheless, it put out tenders
and accepted bids in July 1973 without having reached an agreement on boundaries.[74]

Off-shore Oil Discoveries in Sabah and the Malaysia– Philippine Controversy
•

A new oil producer in Borneo is Sabah where two off-shore producing fields were discovered in 1972 (the Samarang field) and in 1973 (the Tembungo field). In 1975 the Tembungo field was producing about 3,500 barrels per day at a depth of 6,000 feet (1 830 metres). The development of the potential of this area was delayed by protracted negotiations in production-sharing arrangements between the oil companies concerned and the Malaysian government. A controversy on sovereignty over Sabah, although inactive, is legally unresolved at this writing.[75] Although the sovereignty dispute was not in any manner ostensibly linked to a specific economic reason other than property rights, recent developments in the oil sector make the controversy relevant to the present discussion. For background purposes and for an appreciation of the magnitude of the conflict involved, a brief historical summary of the case is given here.

In 1962 the Philippine government formally pressed its claim to political sovereignty over then British North Borneo. This came after the heirs of the Sultan of Sulu approached the Philippine government, having failed in private attempts, to help them 'regain their proprietary rights' over the territory. The details of this case have been exhaustively discussed by various treatises, and will not be dwelt on here.[76] The Philippine claim covers the now independent sovereign state of Sabah, one of the constituent states of the Federation of Malaysia, an area of 29 388 square miles (76 115 square kilometres).[77] Historians indicate that Sabah was ceded to the Sultan of Sulu by the Sultan of Brunei in 1704 in return for help in suppressing a rebellion.[78] Baron von Overbeck was reported to have acquired this territory from the Sultan of Sulu in 1878. The basis for the Philippine claim is that the acquisition was a lease, not a cession, and therefore no transfer of absolute sovereignty by the Sultan of Sulu occurred. Ortiz summarizes the legal arguments in the case in the following manner:

In resumé, on the basis of the historical facts presented, there are solid reasons to sustain, *first*, that the Deed of 1878 was a *lease*; *second*, that even if it was of *cession*, it was null and void as such owing to non-observance of the formalities required and for lack of contractual capacity on the part of Overbeck and Dent; *third*, that the Sultan, although he signed the Treaty of Capitulation of 1878 and constituted himself a loyal subject of Spain, and later, of the United States, remained the sovereign of North Borneo; *fourth*, that the Sultanate was not extinguished nor was the North Borneo territory ever abandoned in a manner that would entitle Great Britain to acquire it by occupation and/or prescription under international law; *fifth*, that, therefore, the successors of Sultan Jamalul Alam since 1878 continue in possession of Borneo; and *sixth*, that therefore, finally, if they cede North Borneo to the Philippine Government as they did sometime last summer [1962], the Philippine Government would then become the rightful sovereign thereover.[79]

The legality of Malaysia's present title to Sabah rests on previous British title. It would appear that this latter is in turn dependent on the validity of the titles held by Baron von Overbeck and Alfred Dent, who received the grants and commissions of 1877 and 1878 from the Sultans of Brunei and of Sulu. Overbeck sold all his rights to Dent, who went on to acquire all the interests in the former American Trading Company of Borneo. Dent later formed a Provisional Association, to which he transferred all his interests and powers in order to acquire a Royal Charter to promote a British company and to develop the North Borneo territories. Ariff, however, concludes that the legal basis of the British (and therefore the Malaysian) claim to Sabah is independent of the validity of the Overbeck-Dent title:

In the Preamble to the Proclamation of the Philippine Independence made by the USA there were specific references to the terms of the Treaties of 10 December 1898 and 7 November 1900 (by which the United States acquired sovereignty over the Philippine Archipelago from Spain) and to the Boundary Convention (concluded with Great Britain and the United States) whereby the line separating the Philippine Archipelago from the State of North Borneo was delimited.

... [The] Constitution of the Republic of the Philippines, ratified in 1947, confirmed the terms of these treaties. It may therefore be assumed that, under the circumstances, the Philippines had adopted the terms of the treaties in their entirety and hence renounced any claims over North Borneo .... Consequently it was left open to the United Kingdom to annex the territory as part of the dominions of the Crown on 15 July 1946, eleven days after the Proclamation of the Philippine Independence.

It may be ... concluded that though British title to the territory of North Borneo arose from the Brunei/Sulu grants and commissions, *it did not depend on them* .... [By] virtue of the Treaty of 1847 concluded with the Sultan of Brunei, the Treaty of 1849 with the Sultanate of Sulu, the Madrid Pro-

tocol of 1885 with Spain, the proclamation of protectorate over the territory in 1888 and the boundary treaties and conventions of 1889 and 1930
with the Netherlands· and the USA, Great Britain acquired an inchoate
title over the territories. The first two treaties established her sphere of
interest and the subsequent treaties and conventions, her sphere of influence
against the other States. The inchoate titles over the territory were gradually
transformed into absolute titles by virtue of the Agreement for the transfer
of the 'Borneo Sovereign Rights' of 26 June 1946 which, ... 'constituted
cession under international law, whilst the North Borneo Cession Order
of 10 July 1946 signified Britain's intention to annex the territories of North
Borneo to form part of the dominions of the Crown.[80]

Some other arguments have tended to focus on the political choice
of the people of Sabah rather than on legal grounds.[81] Proponents
contend that the choice of the people of Sabah to join the Federation
of Malaysia rather than the Philippines supersedes all legal claims.

At the time of this writing, the claim is dormant. At a press conference in Singapore in January 1976, the President of the Philippines stated, in reply to a question, that the Philippines' claim to
Sabah 'is completely at a standstill'. He added, 'We have not pursued
it, nor do we intend to'. He also expressed the hope that ASEAN
would be a vehicle for providing a mechanism for peacefully settling
border disputes involving territory.[82] In July 1976, he was reported
to have stated that the Philippine Constitution would not allow him
to give up the claim based on historic grounds.[83] In July 1977, however, there were hints that the possibility of dropping the claim
would be discussed by the top leaders of the two nations.[84] On open-
ing day of the Second ASEAN Summit in early August 1977, the
Philippines President announced that his government was ready
to drop its claim to Sabah,[85] and clarifications on what this meant
followed. The statements did not, however, categorically indicate
that the claim would be dropped, and much ground work remained
to be done before the legal aspects could be clarified. The latter
included, among other things, an amendment to the Philippine
Constitution.

Even if the legal question were eventually decided in favour of
the Philippines, the question of Sabah's political status would still
remain to be settled. The problem has serious political as well as
economic implications, discussion of which is beyond the scope of

this study. Suffice it to say that the petroleum resources of Sabah have certainly made the area economically more attractive to both states involved in the controversy.

1. *Petroleum News S.E.A.*, June 1973, p. 20.

2. *Petroleum News S.E.A.*, February 1974, pp. 25-8.

3. *PN*, June 1973, p. 28.

4. United Nations, ECAFE, *Proceedings of the Fourth Symposium on the Development of Petroleum Resources of Asia and the Far East* (New York, 1972), Vol. 1, p. 178.

5. *Petroleum News S.E.A.*, May 1973, p. 21.

6. Thailand, Ministry of Industry, Department of Mineral Resources, *Petroleum Activities in Thailand* (Bangkok: March 1976), pp. 1-2.

7. *Petroleum Activities in Thailand*, pp. 1-2.

8. *Petroleum News S.E.A.*, July 1973, p. 31.

9. *Petroleum News S.E.A.*, November 1972, p. 6.

10. Exxon acquired a 35 per cent interest in Elf's acreage in late 1972. *Petroleum News S.E.A.*, September 1972, p. 7.

11. *Petroleum Press Service*, August 1973, p. 311.

12. *PN*, February 1974, p. 28.

13. *PN*, June 1973, p. 20.

14. Declaration du Gouvernement Royal en Date du 27 Septembre 1969 relative a la mer territoriale et au plateau continental du Cambodge.

15. *PN*, June 1973, p. 20.

16. Ibid.

17. *Petroleum News S.E.A.*, June 1975, p. 20.

18. *Petroleum News S.E.A.*, 30 September 1974, p. 3.

19. *PN*, June 1973, p. 20.

20. See remarks attributed to Professor Shigeru Oda in UNDP, CCOP, *Report of the Sixth Session*, 1969, p. 18.

21. *Petroleum News S.E.A.*, October 1974, p. 7, and earlier discussion on Koh Kut Island.

22. *Petroleum Press Service*, August 1973, p. 314.

23. *Petroleum News S.E.A.*, 30 June 1974, p. 9.

24. *Petroleum News S.E.A.*, February 1974, p. 25.

25. J.K. Blake, 'Promising concession prospects in South Vietnam', in *Petroleum News S.E.A.*, February/March 1973, p. 18.

26. *PN*, February 1974, p. 28.

27. *PN*, 30 June 1974, p. 9.

28. *The Petroleum Economist*, October 1974, p. 391.

29. *PN*, 30 September 1974, p. 3.

30. Far Eastern Economic Review, *Asia 1976 Yearbook* (Hong Kong: December 1975), p. 83.

31. The term 'China' is used here to refer to mainland China and its government whose seat is in Peking. The term 'Taiwan' is used to refer to the Chinese Nationalist government in Taiwan whose seat is in Taipei. For clarity, especially where the actions of specific governments are discussed, the following terms will be used interchangeably: 'Peking' for 'China'; 'Taipei' for 'Taiwan'; 'Saigon' for 'South Vietnam'; and 'Hanoi' for 'North Vietnam'. (See footnote 2 of Introduction.)

32. The data under this heading are a conglomerate of information largely culled from several sources: Gerard Corr, 'The world', *Straits Times* (Singapore), 30 November 1975, p. 12; David Rogers, 'Oil wealth sparks a dispute over the Spratlys', *Straits Times*, 6 July 1976, p. 8; Michael Richardson, 'The fuse that could set off an international explosion ...', *Straits Times*, 19 July 1976, p. 16; Bernard Wideman, 'Manila probes a sensitive spot', *Far Eastern Economic Review*, 28 May 1976, p. 115; *Bulletin Today* (Manila), 6 May 1976, p. 3, 'No easy solution seen on Spratly, Paracel problem'; *Economic Bulletin* (Singapore), July 1975, p. 26; *New Philippines* (Manila), February 1974, pp. 6-9, 'Government states position on imbroglio over islets'; *New Nation* (Singapore), 26 July 1976, p. 7, 'Vietnam stamps its claim'; *Petroleum Economist* (March 1976), p. 114, 'Philippines', and July 1976, p. 279, 'East Asia offshore: claims and counter-claims'; *Petroleum News S.E.A.*, February 1974, pp. 32-3, 35, 50, 'Rough water in the South China Sea'; *South China Morning Post* (Hong Kong), 16 June 1976, 'Islands of multiple claims'; *Straits Times*, 16 June 1976, p. 2, 'Peking warns: hands off Spratly isles'; *Straits Times*, 18 June 1976, p. 2, 'Spratly isles: drilling goes on despite China warning'; *Straits Times*, 19 June 1976, p. 3, 'Philippines can defend Spratly interests', *Straits Times*, 21 June 1976, p. 4, 'Romulo for China talks on Spratly?'; *Straits Times*, 22 July 1976, p. 4, 'Manila ready to renew claim'; U.S., House of Representatives, Committee on Foreign Affairs, *Oil and Asian Rivals* (Washington, D.C.: U.S. Government Printing Office, 1974). Additional materials are cited for more specific identification where needed.

33. The term *abyssal plain* is used to refer to the deep ocean floor (an average depth of 4 000 metres).

34. Lee Lai-to, 'The PRC and the South China Sea', *Current Scene*, Vol. XV, No. 2 (February 1977), p. 5.

35. *New Philippines*, p. 7, and *PN*, February 1974, p. 36.

36. For a more detailed discussion of China's claim, see Lee Lai-to, 'The PRC and the South China Sea', op. cit., pp. 1-12.

37. See Corr, op. cit.

38. *PN*, February 1974, p. 36.

39. Lee Lai-to, op. cit., p. 5.

40. Information in this paragraph from Lee Lai-to, op. cit., p. 6.

41. Arthur M. Schlesinger, Jr., *A Documentary History of U.S. Foreign Policy 1945-1973* (Vol. 4, *The Far East*, p. 55), Chelsea House, 1973, cited by Lee Lai-to.

42. See section on the Philippine claim below.

43. See section on other South China Sea islands below.

44. See Lee Lai-to, op. cit., p. 4.

45. Ibid., p. 4.

46. *South China Morning Post*, 16 June 1976, 'Islands of multiple claims'.

47. The term *res nullius* is defined in Webster's Third International Dictionary (Unabridged) as 'a property belonging to no one whether because never appropriated ... or because abandoned by its owner but acquirable by appropriation.'

48. It at one time used the 200-mile economic zone concept but later consistently used the continental shelf concept, presumably because the economic zone was not legally existent.

49. *New Philippines*, p. 11.

50. *South China Morning Post*, 16 June 1976.

51. *Straits Times*, 16 June 1976.

52. *Straits Times*, 19 June 1976.

53. As of mid-1977, this date had not been reset.

54. *Far Eastern Economic Review*, 28 May 1976, p. 115.

55. *Daily Express* (Manila), 20 July 1976, pp. 1 and 2, 'Traces of oil found in Palawan drill site'.

56. See *Straits Times*, 19 June 1976, and Richardson, 19 July 1976.

57. Richardson's title.

58. *Petroleum News S.E.A.*, October 1976, p. 4, 'General News'.

59. See *Petroleum Economist*, September 1976, p. 362.

60. See pages 60 and 61 above.

61. *Convention on the Continental Shelf*, Article 6.1.

62. See *Petroleum News S.E.A.*, 30 November 1974, p. 7.

63. Lee Lai-to, op. cit., p. 9.

64. Lee Lai-to, op. cit., p. 5.

65. *PN*, February 1974, p. 36.

66. Lee Lai-to, op. cit., p. 5.

67. Ibid., p. 7.

68. Ibid.

69. *PN*, February 1974, p. 32.

70. See Lee Lai-to, op. cit., p. 8.

71. *Oil and Asian Rivals*, p. 110. Lee Lai-to (op. cit., p. 4) reports that Indonesia supported China's claims over the Spratlys and the Paracels. His supporting statement, however, referred only to Indonesia's interpretation that the San Francisco Peace Treaty of 1951 had stated that the Paracel Island belonged to China.

72. *Asian Wall Street Journal* (Hong Kong), 9 September 1976, p. 1, 'Dispute over islands in South China Sea is fueled by oil find'.

73. J.K. Blake, op. cit.

74. *Petroleum Press Service*, August 1973, pp. 310-11.

75. The Philippines' President announced that the Philippines was 'taking steps to eliminate one of the burdens of ASEAN—the claim of the Philippine Republic to Sabah'. See *Straits Times* (Singapore), 5 August 1977, p. 1. By mid-October 1977, however, there were no reports that steps toward such relinquishment had formally begun on a government-to-government level.

76. Some excellent references are: M.O. Ariff, *The Philippines' Claim to Sabah* (Kuala Lumpur: Oxford University Press, 1970); Pacifico Ortiz, S.J., *Legal Aspects of the North Borneo Question* (Manila: Bureau of Printing, 1964); Philippine Government, *Philippine Claim to North Borneo*, Vols. I and II (Manila: Bureau of Printing, 1963); Philippine Government, 'Philippine policy statement in the United Nations', delivered before General Assembly, 23rd session, New York, 15 October 1968; Michael Leifer, *The Philippine Claim to Sabah* (Centre for South-East Asian Studies, University of Hull, 1968); and S. Jayakumar, 'The Philippine claim to Sabah and international law', *Malaya Law Review*, Vol. 10, No. 2 (1968), pp. 306-35.

77. Ariff, op. cit., p. 1.

78. Ortiz, op. cit., p. 6. The rebellion was dated 1662 to 1675 by James P. Ongkili, *The Sunday Mail* (Kuala Lumpur), 15 July 1973.

79. Ortiz, op. cit., pp. 39-40.

80. Ariff, op. cit., pp. 24-5.

81. See, for example, the articles by James P. Ongkili in *The Sunday Mail* (Kuala Lumpur), 8 and 15 July 1973; and the address by the Prime Minister of Malaysia in Parliament on 15 October 1968, 'Philippine Grab Bill: Unprecedented in international relations' (Kuala Lumpur: Department of Information).

82. *Straits Times* (Singapore), 29 January 1976, p. 12.

83. *Petroleum News S.E.A.*, September 1976 News Supplement centrefold map.

84. *Straits Times* (Singapore), 16 July 1977, p. 3, 'Philippines to drop claim to Sabah?'.

85. *Straits Times*, 5 August 1977, p. 1, 'Summit surprise by Marcos'.

V

Potential Conflict Situations arising from Geological Settings and Environmental Phenomena

THE development of petroleum resources involves activities which impact across man-made boundaries. Settlement of boundary disputes may not completely end petroleum-related international disputes in the areas of exploratory research, petroleum production, and environmental control. The natural phenomenon of petroleum accumulation underneath the seabed respects no political or juridical surface boundaries. In the same way, whatever limits are agreed upon among nations, the sea and air retain the characteristics of an indivisible common property—it is impossible to restrict the movement of the waters or the wind within predetermined legal boundaries. Therefore, property rights disputes are bound to arise in the process of exploring for and producing off-shore petroleum, as well as in efforts to protect the coastal environment from damage.

Geology-Related Potentials for Conflict

To understand how property rights can be disputed in the process of exploring for and producing petroleum even when territorial boundaries have already been agreed upon, it is necessary to understand the basic technological and economic aspects of producing this resource. Thus, this section will touch on the rudiments of oil accumulation and extraction, and then the causes for property rights disagreements.

RUDIMENTS OF OIL ACCUMULATION AND PRODUCTION[1]
Oil and natural gas are fossil fuels occurring naturally in certain sedimentary basins and geosynclines. Petroleum forms in sedimentary rocks wherein organic matter buried with the sediments becomes fossil fuels. In the course of time coal, oil, or gas were formed by

the combined action of pressure, temperature and physical-chemical processes on dead marine animal and plant debris.

Not all sedimentary rocks are favourable for the occurrence of petroleum, and so petroleum accumulations are distributed unevenly throughout the world. To date a large proportion of the oil and gas discovered are located in a few areas. Large deposits of petroleum require the presence of, or proximity to, thick sedimentary rock strata that were deposited in an appropriate marine environment and suitable geological traps. Fossil fuel is produced through chemical change in biological material that was deposited during the last 600 million years in thick layers of sediments. As other layers of sediment rocks piled over time on top of the original sediment containing hydrocarbons, the pressure forced the petroleum to 'migrate' out of the 'source' rocks until some escaped into the atmosphere or was trapped by an impervious layer, or 'cap' rock. Thus the explorer is really not merely looking for petroleum, but for possible petroleum 'traps' where the migrating oil finally accumulated; this 'reservoir' rock is a porous layer (usually sandstone or limestone). Several factors affect reservoir conditions. Some of these are: (1) the ability of the rock formation to hold fluids, or its porosity; (2) the ability of the rock to have fluids transmitted through it, or its permeability; and (3) the pressure in the reservoir.

The presence of a petroleum reservoir can be determined only by drilling into the trap. Oil production is a displacement process, with gas or water, or both, filling up the portion of the reservoir vacated by oil. Oil is produced by puncturing the impermeable cap of the reservoir and creating a lower pressure in the well bore than exists in the rest of the rock formation. If the pressure differential is sufficient, the oil or gas will flow through the well bore to the surface. If oil or gas will not flow naturally, the necessary pressure differential can be created by pumping. Gas, water, or some other substance may be used to supplement and maintain the natural drive in a reservoir.

As a general rule, some gas is always found with oil, but quite frequently gas is also found without oil. Gas, an extremely important fuel by itself, is also of critical importance in the production of oil, because it is one of the major natural driving forces in an oil reservoir. Under some reservoir conditions, natural gas is dissolved in the crude oil which, when brought to the surface, releases the gas.

In the development of a mineral resource that is depletable, not all the resource discovered is recoverable for use, for technological or economic reasons.[2] The recoverable portion of the total volume of oil in a reservoir depends on a number of factors. A primary determinant is the source of the reservoir's natural drive. Recovery factors differ from field to field,[3] with variations arising from the differences in the reservoir environment in which the hydrocarbons occur. In addition, the rate of extraction can affect total ultimate recovery. The following quotation explains this:

> The requirements for efficient recovery of the oil from a reservoir are not taken care of by chance; they may be fulfilled only through careful and deliberate action by the producers. Experience has shown that one of the most essential factors in meeting these requirements is control of the rate of production. Excessive rates of withdrawal lead to rapid decline of reservoir pressure, to release of dissolved gas, to irregularity of the boundary between invaded and non-invaded sections of the reservoir, to dissipation of gas and water, to trapping and by-passing of oil, and, in extreme cases, to complete loss of demarcation between the invaded and non-invaded portions of the reservoir, with dominance of the entire recovery by inefficient dissolved-gas drive. Each of these effects of excessive withdrawal rates reduces the ultimate recovery of oil.
>
> ... there has developed the concept of the *maximum efficient rate* of production, often referred to as the M.E.R. For each particular reservoir, it is the rate which if exceeded would lead to avoidable underground waste through loss of ultimate recovery.[4]

PROPERTY RIGHTS DISPUTES ARISING IN THE
EXPLORATORY STAGE

With an understanding of the basic conditions under which petroleum is accumulated and to which search must be directed, it is easy to appreciate that the initial research into geological conditions favouring such an accumulation cannot be site-specific in a narrow sense. Geological and geophysical research will tend to be regional at the early stage, on the basis of which these scientists would then recommend acreage acquisition and exploratory drilling on the probability that (1) hydrocarbons exist in a specific geographic area; (2)

that these hydrocarbon deposits can be found and produced; and (3) that the deposits found will be economic to extract and sell.

It is because of the desirability of a regional investigation at the initial stage that some concern was expressed at the U.N. debates over freedom of scientific research or investigation on the continental shelf or economic zone. After all, initial geological and geophysical surveys in petroleum exploration, prior to identification of specific areas where more detailed work would be conducted and prior to acquisition of acreage, are in fact research activities. As pointed out by a legal consultant of the UNDP/CCOP, some difficulty could arise in arriving at a precise distinction between research concerning the continental shelf and exploration of the continental shelf, each of which is covered by differing legal provisions on national jurisdiction under international law.[5] Under the provisions of the 1958 Continental Shelf Convention, the coastal state has exclusive right to explore its shelf. At the same time the Convention provides that, while consent of the coastal state was required prior to any research on its shelf, that state should not normally withhold consent to requests made by a qualified institution for 'purely scientific research' into the physical or biological characteristics of the shelf (Article 5.8).

In discussions in the Third Law of the Sea Conference, most nations regarded scientific research conducted in marine areas within a coastal state's territorial and patrimonial sea as falling under the state's control and jurisdiction. Thus, more restrictive but potentially conflicting provisions similar to the above appeared in the Revised Text.

Articles 53 to 56 of Part III of the Revised Text (Articles 239-245 in the Composite Text) provided for the promotion of international cooperation in marine scientific research and in the flow of scientific data and information resulting from such research. Article 61 provided, on the other hand, that the results of research related to the exploration and exploitation of the resources of the economic zone and continental shelf of a state may not be 'published or made internationally available against the express wish of that state'. The distinction might be defined as follows: the results of scholarly studies done by non-profit institutions are freely publishable; those of profit-making organizations for the purpose of profit-taking activities are proprietary information belonging to the coastal state.

Thus, the published results of a regional study of off-shore mineral resources (including petroleum) in South-East Asia undertaken in the late 1960s by the United Nations' special committee, CCOP, served to attract private investment into the region. On the other hand, exploratory voyages of a Turkish seismic research ship in the Aegean Sea gave rise to objections from Greece. Greece considered the Turkish survey a violation of its rights under the Geneva Convention on the Continental Shelf because, it complained, Turkey's ship was *exploring* on the continental shelf of Greek islands in the Aegean. Turkish officials were said to consider the ship's research a *scientific survey*, without its necessarily constituting a claim to ownership of the seabed. At the same time Turkey was said to reject the notion that it had to inform Greece of its survey; it claimed the islands did not generate their own continental shelf but were mere protuberances on the continental shelf of the Turkish mainland. The International Court of Justice was reported to have rejected a Greek application for a temporary ban on Turkish surveying in the disputed waters.[6]

The difficulties that would have arisen with Article 61 appeared to have been eliminated in the revised provisions on dissemination of information and knowledge in the Composite Text (see Article 245).

Within the South-East Asian region, there is a consensus with respect to the coastal state's jurisdiction over scientific research in its adjacent waters but the desired breadth of such jurisdiction varies by state. The minimum limit of jurisdiction within the territorial waters appears to be unanimously accepted, but some countries would like to see that jurisdiction extended to the economic zone or continental shelf, whichever is wider. (See Table 3.1 again.)

The Philippines seeks jurisdiction at least within its archipelagic waters,[7] and Thailand supports jurisdiction within its territorial waters for security reasons.[8] Thailand further expressed the belief that the benefits of scientific research could be shared by the international community, especially the developing coastal states in whose jurisdiction areas the research would be conducted. Burma would require prior consent of the coastal state before any research is conducted on and about its continental shelf.[9]

Indonesia, Malaysia, and Singapore claim jurisdiction over scientific research in the economic zone.[10] In addition, Singapore and other developing states proposed that scientific research in extra-territorial waters be controlled by an international authority. Singapore, therefore, viewed with alarm a proposal by some developed and developing countries for freedom to conduct scientific research in extra-territorial waters.[11]

Regional cooperation in such research is, of course, possible. Five South-East Asian nations with membership in the UNDP/CCOP were reported to have agreed to undertake a study of proposals for massive off-shore oil exploration and development projects on a regional cooperative venture basis.[12] These countries are Indonesia, Malaysia, the Philippines, Singapore, and Thailand, which also compose the Association of South-East Asian Nations (ASEAN). In the Declaration on the establishment of the ASEAN Council on Petroleum (ASCOPE), these nations specified as one of ASCOPE's aims 'collaboration and mutual assistance in the development of the petroleum resources in the region' as well as cooperation and assistance in research.[13]

DISAGREEMENTS ARISING FROM AN OIL POOL SHARED BY TWO OR MORE NATIONS

In the production stage, even when surface boundary limits have been agreed upon, property rights disputes may still arise when petroleum accumulation is spread out under two or more national jurisdictions. Just as there is no way of assuring that a reservoir discovered by an operator falls neatly within the confines of his acreage, there is no guarantee that such a discovery would be co-terminous with the national boundaries of the host country if the discovery is close to the border. There is always the possibility of leases straddling a field and a similar likelihood for national boundaries to do the same (see Fig. 8). The conflicts could arise from two sources: (1) accusations that the neighbouring country's operator is draining oil from one's acreage; or (2) should there be a unitized operation agreement between the neighbouring countries, disagree-

108

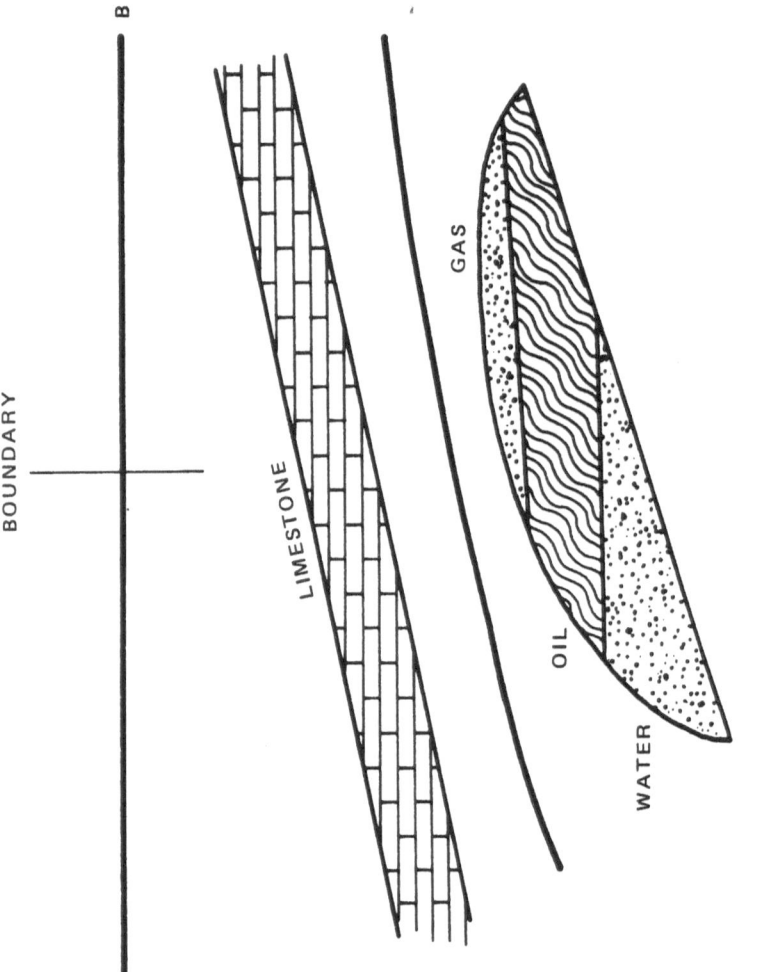

Fig. 8. Stratigraphic Wedge Bisected by National Boundaries.

ment on the optimum rate of production that would assure maximum ultimate recovery from the reservoir.

Before discussing the need for unitization of operation, it would be useful to describe a relevant concept, the 'law of capture', when several owners own one oil pool, and the experience in the United States.[14] The 'law of capture' in the system of oil property rights basically states that the owner of acreage which produces oil from a well on that acreage is recognized as having produced or 'captured' the oil, even though it is possible that a part of it has been drained from the sub-surface of contiguous property. Where there are several owners drilling from a common pool of oil, the rational behaviour of each, in the absence of any agreement or regulation, is to extract as much and as fast as possible in order to avoid losing out to one's neighbour. The producer who moderates his current flow risks allowing other producers to pump oil he could have had. Since each producer in the pool knows this, all may pump too rapidly. Thus, prior to field regulation in the United States, the result was dense drilling in an ever widening pattern, to prevent migration of oil to contiguous areas.

The same problem could crop up in South-East Asia's off-shore areas. There are, in fact, unitized operations in Indonesia.[15] It is conceivable that the problem may arise at the boundaries of national jurisdictions instead of only between leases, so that there would be a need to unitize operations across territorial limits. In that case the problem would no longer be limited to an agreement between private parties on the rate of production; there would arise a need to agree on such rates at state levels.

The difficulty of identifying, keeping track of, and asserting property rights over a part of a common pool creates a problem of efficient production rates in terms of the long-run interests of society. A producer will tend to neglect the likely effects of his actions on the availability of the resource in the future, if he believes that his neighbour is pumping out some of the oil in his acreage and that he cannot hope to reap the future benefits from foregoing some current profit. 'Pooling' or 'unitization'[16] would be the answer to precluding wasteful rates of production or accusations of pumping out the oil from a contiguous area, whether such accusations are true or not. It is not an easy answer to come by, however. The agreement would have to answer the question: what is the long-term optimum ('socially economical') amount of oil that should be produced under any given set of technical and economic conditions? This would involve setting up regulations that would assure the proper use of the natural driving forces of a reservoir and agreeing on formulas that would provide for equitable sharing of the contents of the pool among those who have the right to produce. This may pose problems. When there is more than one owner, there is always the problem of determining the optimum production rate that would serve the interests of those involved. On the state level, there could be diversities in goals or social priorities that could result in differences in each party's desired rate of production. States could also differ on the optimum time distribution of indigenous petroleum resources or on the social discount rate to be applied to such resources. The problem of sharing is further complicated by the fact that, at the time of unitization (usually early in the life of the field), detailed knowledge of the reservoirs and of the amounts recoverable under each tract is not yet available. There are, of course, ways of handling this through adjustments at a later date, but there is still the possibility of protracted bickering and lengthy negotiations before arriving at a mutually acceptable arrangement.

The possibility that conflicts of this type would arise is not minor. The Straits of Malacca are again a good example of an area where such problems could arise, since the Straits may not be totally devoid of commercially exploitable petroleum resources. Even excluding the Straits, any place on the continental shelves of East Asia could be a source of conflict, whether this be on the Andaman Sea or the Gulf of Thailand. As will be recalled there was real concern in 1974 when oil and gas shows in a concession granted by Thailand were located within 25 kilometres of a disputed border with Cambodia and South Vietnam.[17]

Commonality of Water Resources and the Conservation/Pollution Problem

Every aspect of the petroleum industry involves contact with the

environment. Inevitably, environmental externalities[18] in exploration, production, transportation, and marketing of petroleum are
bound to be transmitted across international political boundaries via
two primary mechanisms—natural water flows and atmospheric
motion. Our concern here is the effect on inter-country relations of
damage to the environment in the course of developing South-East
Asia's petroleum resources. Ingo Walter categorizes 'transfrontier
pollution' in terms of direction of impact: *one-way*, where pollution
originating in one country damages one or more other countries, but
without causing significant damage in the originating country, and
two-way or reciprocal, where pollution significantly damages both
the country originating it and others.[19]

Marine pollution can emanate from several sources including the
following: oil spills in drilling operations or tanker accidents, dumping of drilling mud into the sea during off-shore operations, tanker
washings in international waters, silting from dredging operations,
and chemical pollution. Of marine pollutants, oil is the most common. Estimates of the volume of oil pollutants in the seas range
from 1 million metric tons to 1.5 million metric tons per year, the
latter covering all forms of marine oil pollution including that from
off-shore drilling.[20]

While it is possible to define the limits of territorial waters or the
economic zone legally, the sea by nature raises a problem with regard to any nation's ability to contain, within its boundaries, adverse effects of exploration and producing activity on the environment. Ordinary exploration and producing activity may drive away
schools of fish or disturb the nearby habitats of shellfish and crusta-
ceans. If there are any blowouts that cause oil spills or if spills occur
during transport of the produced oil, such spills may spread to adjoining waters and damage living resources on the water or on the
adjacent coasts. Marine oil pollutants destroy fish eggs as well as
sea birds. In Asia, especially, development of the petroleum industry is seen as a potential hazard to one of its most significant protein sources—its fish-life.

This problem of trans-frontier pollution could arise especially if
exploration/producing activity is conducted in national waters close
to the boundaries of another state. Since it is not possible to put a

fence around jurisdictional waters to contain any adverse effects from such activity, this type of 'common-pool' problem could be an irritant if the adjacent state is not similarly engaged in exploration/ production. This type of irritation has, in fact, already occurred in connection with oil tanker passage in the Malacca Straits, where Malaysia is reported to have complained that, while it is concerned about pollution and the environment, Indonesia is more concerned about the 'power problem'.[21]

Some of the marine accidents related to oil exploration in South-East Asia are briefly described in the book by Howell & Morrow, *Asia, Oil Politics and the Energy Crisis.* In the Java Sea in late November 1970, the drilling rig *Milton G. Hulmes* lost control of shallow, non-flammable gas. A little over two months later, on 6 January 1971, another drilling rig, the *Big John*, lost control of shallow gas off Sarawak when the gas vented under the rig and caused it to sink partially before catching fire. Another accident occurred in September 1972, when the drilling rig *Transworld 60* had a blowout 35 miles (21 kilometres) off Burma's Irrawaddy River delta; gas bubbled from the hole for months. On the North-West Shelf of Australia, the *Sedco 135 G* lost control of gas flow. The hole blew for over two years, and released billions of cubic feet of natural gas into the surrounding area before the flow subsided.

Towards the end of 1973, it was reported that there had been at least fifteen serious blowouts in off-shore exploration in Asia. As all of those blowouts were related to gas rather than oil wells, the damage to the sea environment was relatively lower than that resulting from óil well blowouts.[22] (There was, of course, substantial loss of energy resources.)

The UNDP/CCOP has promoted the development of anti-pollution legislation and regulation in member countries. At the eleventh session of the Committee, delegates from Malaysia and Singapore reported that ministries on the environment had been established by their respective governments. These ministries were concerned with pollution and environmental protection problems, particularly those related to marine pollution.[23]

At the same session the Indonesian delegation reported that several laws had been passed since 1970 against marine pollution from oil spills and leaks. In 1972 the existing law was extended to cover not only the continental shelf areas but all of Indonesian waters. More stringent laws were passed by the Ministry of Mining in 1973. Regulations cover prohibitions against oil spills, against dumping of drilling mud into the sea before it is neutralized, and against releasing gas into the atmosphere before it is flared. These anti-pollution measures and safeguards imposed on oil companies operating in Indonesian waters are enforced by a government team. At the time of reporting four major oil companies had complete marine anti-pollution equipment.

At the Third Law of the Sea Conference South-East Asian countries have generally argued for coastal state control over marine environmental standards within territorial waters or economic zones. (See Table 3.1 again.)

Indonesia, Malaysia, and the Philippines supported the position that the coastal state should have sovereign rights over protection of its marine environment within the economic zone.[24] The Philippines co-sponsored draft articles for control within the economic zone, Article 7 of which provided that the coastal state should have the 'right and duties' to preserve and protect the marine environment and to prevent and control pollution.[25] Such controls would be in accordance with 'internationally agreed rules'.

Thailand, situated in a semi-enclosed sea traversed by international navigation, has a special interest in the prevention and control of pollution. It subscribes to the idea of an international standard for the sea, the seabed, and the subsoil, especially with regard to liability, remedy and compensation for damages.[26]

Possibilities of trans-national cooperation exist. Part XII of the Composite Text provided for global and regional cooperation by states to protect and preserve the marine environment. This includes cooperation in related scientific research programmes, and exchange of information and data so acquired.[27] And on 25 February 1977, in an agreement to safeguard the safety of navigation in the Straits of Malacca and Singapore, the bordering countries of Indonesia, Malaysia and Singapore agreed to formulate a joint policy to deal with marine pollution.[28]

1. The discussion in this section will draw heavily from Bernardo F. Grossling, *Latin America's Petroleum Prospects in the Energy Crisis*, U.S. Geological Survey Bulletin 1411 (Washington, D.C.: U.S. Government Printing Office, 1975), pp. 10-25; Wallace F. Lovejoy and Paul T. Homan, *Economic Aspects of Oil Conservation Regulation* (Baltimore: The Johns Hopkins Press, 1967), pp. 59-61; U.S., Council on Environmental Quality, *OCS Oil and Gas—An Environmental Assessment* (Washington, D.C.: U.S. Government Printing Office, 1974), p. 18; and R.C. Kuller and R.G. Cummings, 'An economic model of production and investment for petroleum reservoirs', *American Economic Review*, Vol. 64 (1974), pp. 66-79; 'A primer on petroleum origin and exploration' in Oriental Petroleum and Mineral Corporation, *Annual Report* (Manila, 1970).

2. The notion of *recovery efficiency*, expressed in percentage terms, refers to the quotient of the total proved and recoverable petroleum at a point in time divided by the estimated cumulative total in reservoirs known to exist at that point in time. See Corazon M. Siddayao, 'Potential supply of petroleum from tertiary recovery', in 'Patterns in the utilization of the major energy resources of the United States', unpublished report for the Ford Foundation, 1974.

3. A *field* is the general area underlain by one or more reservoirs formed by a common structural or stratigraphic feature. R.E. Megill, *Exploration Economics* (Tulsa, Oklahoma: The Petroleum Publishing Company, 1971), p. 12.

4. Stuart E. Buckley, editor, *Petroleum Conservation* (New York: American Institute of Mining and Metallurgical Engineers, 1957), pp. 151-2, quoted in Lovejoy and Homan, op. cit., pp. 205-6.

5. See U.N., ECAFE, CCOP, *Report of the Sixth Session*, 1969, p. 17.

6. See *Guardian Weekly* (Paris), 1 August 1976, reprinted as 'Dispute over Aegean seabed' in *The Mirror* (Singapore), 16 August 1976; and *Asian Wall Street Journal*, 13 September 1976, p. 6.

7. *Official Records*, Vol. III, pp. 226-7, and Document A/Conf. 62/C.2/L.49.

8. *Official Records*, Vol. I, p. 148.

9. *Official Records*, Vol. II, p. 224, and Law No. 3 of 1977.

10. See *Official Records*, Vol. II, pp. 198, 207-8, 268; and *Straits Times*, 14 July 1975.

11. See *Straits Times*, 28 August 1974.

12. *Singapore Trade and Industry*, Supplement, February 1976, p. 19.

13. See Declaration on the Establishment of ASEAN Council on Petroleum (ASCOPE) reproduced in CCOP *Newsletter*, September 1976, pp. 13-14.

14. The following discussion will be based largely on Lovejoy and Homan, op. cit., pp. 21-3, 49, 67, 77-8. The 'law of capture' was established in court decisions in various states in the United States before the importance of underground migration had been completely understood in the oil industry. See Dennis Epple, *Petroleum Discoveries and Government Policy* (Cambridge, Mass.: Ballinger Publishing Co., 1975).

15. Conversation with Allen Hatley.

16. The combination of leases or other acreage interests so as to vitiate the boundary between leases is referred to as 'pooling'. When the combination involves a number of leases covering a reservoir, the term 'unitization' is usually applied. See Siddayao, op. cit., 'Tertiary recovery'.

17. See *PN*, 30 June 1974, p. 9.

18. The term *externalities* refers to discrepancies between social and private costs or benefits. That is, it is the externally produced benefit or cost to society that is generated by private production or consumption.

19. Ingo Walter, *International Economics of Pollution* (London: The MacMillan Press, Ltd., 1975), p. 141.

20. Howell and Morrow, op. cit., p. 116.

21. *Far Eastern Economic Review*, 27 August 1976, p. 59.

22. The above data are from Michael Morrow, 'Blowout at Houston firm's rig leaves nuclear-like crater off Burma', Release No. 485, Dispatch News Service International, December 1972, reported in Howell and Morrow, op. cit., p. 117.

23. ESCAP, CCOP, *Report of the Eleventh Session*, 1974, pp. 36-8.

24. *Official Records*, Vol. II, pp. 198, 207-8, 249.

25. Document A/Conf. 62/C.3/L.6.

26. *Official Records*, Vol. I, p. 148.

27. See *Composite Text*, Articles 198-202.

28. *Straits Times*, 25 February 1977, p. 11, 'Straits: Joint policy on pollution'.

PART THREE

Economic and Policy Implications

VI

Some Economic Factors bearing on the Resolution of Conflicts over Oil Resources

REGARDLESS of the economic or political system existing in a country, the economic objective of the state is to achieve maximum economic welfare, or maximum satisfaction in the utilization of its resources. This implies the distribution of goods in a manner that would maximize total satisfaction. But to attain this end, it must seek to allocate in an efficient manner given total flows of productive services per unit of time to different products and processes. In relation to petroleum resources, the maximization of welfare may be viewed in terms of their development to increase domestic supply at minimum social cost, and also maximization through increased satisfaction in consumption over time.

A pool of oil is a capital asset to society or its owner, even though it is not reproducible and the existing stock can only decrease through time. Ideally society wants to obtain the maximum net benefits from this resource over time. But a resource has economic value only if it can be produced and marketed. In South-East Asia, the economic value of off-shore petroleum depends not only on how much there is, but also on whether or not it can be consumed. Costs of delays in its discovery, extraction, and consumption detract from the value to society of the resources now hidden in the seabed. Resources available in ten years may not be as valuable as resources desired and available now, and so their value has to be discounted.

The value placed on petroleum as an input to economic growth (which is often taken as a surrogate measure of economic welfare) implies that optimum access is desirable. The growing attractiveness of developing indigenous resources, because of the high costs of imports, underscores the desirability of access to such resources in usable form. Given the existence of indigenous resources in South-East Asia, advances in technology that permit production of off-shore resources, and world prices that make domestic resources cost-competitive, impediments to access to such resources no doubt detract from or reduce the opportunities to increase national welfare. Therefore, actual and potential property rights disputes over such resources, by inhibiting access, are impediments to maximizing opportunities for increasing national economic welfare. A rational approach to maximizing benefits would, therefore, be to seek agreement where conflicts exist.

To appreciate more fully the economic considerations involved, it would be useful to understand (1) the basic technological and economic problems of petroleum resource development; (2) the capital expenditure levels involved in off-shore development; and (3) some major economic and environmental impacts of such development. The reader is reminded at this point that the scope of this study does not involve an actual industry data research in South-East Asia nor empirical estimates of possible impacts on each country that might be affected. Rather, it seeks to identify the economic variables, given experience elsewhere, that are likely to be affected by the development of petroleum resources, on the one hand, and by the impediments to such development, on the other. Such recognition is a necessary first step to future in-depth policy analyses or planning.

The Economics of Petroleum Resource Development

BASIC ECONOMIC AND TECHNOLOGICAL CONDITIONS IN THE DEVELOPMENT OF PETROLEUM RESOURCES

The mere physical presence of oil and gas resources does not guarantee that they will be produced. Certain technical and economic conditions must favour exploration and development[1] of those resources. In the search for petroleum, however, decisions must be made to spend huge sums of money before all the facts are known. Therefore, the third and most important single variable in the assessment of an exploratory investment is risk.[2] Modifying Megill's formula slightly, the economic decision in exploration may be expressed simplistically in equation form as follows:

Hydrocarbon potential + technology + economics

+ risk = decision[3]

This equation states that the variables include identification of the
potential of finding oil or gas resulting from geological and geophysical studies, the technology available to discover and later to produce
this potential, the costs and gains from exploring further and later
developing the potential reservoir, and the probability of losing out
on the venture. Legal and administrative considerations would
normally be included in the economics variable, but in a country
where it is difficult to predict changes in the institutional framework within which the producer must operate, the legal/administra-
tive factor may be included in the risk element.

1. *Technical consideration.*[4] When any portion of a land or shelf
area is to be developed, the area in question is divided into tracts
which are leased out to prospective developers (e.g., oil companies)
under some form of contract. Before such contracts are entered
into, however, preliminary geological and geophysical surveys will
already have been made, and the area's sedimentary basins will have
been evaluated for potential petroleum accumulation.

In addition to the potential presence of favourable geological
conditions, the technology and infrastructure to conduct exploration
and development activities in the area of interest should be available.
In the case of off-shore development, the technology to drill and produce at the water depths necessary should be existent.

2. *Economic considerations.* Basic economic factors also determine
whether or not the area to be developed is a viable prospect, since
the producer must calculate his return on investments in the short as
well as in the long term. Even before a well is drilled, a producer
must, therefore, consider the price structure in the prospective market, the distance to consuming areas or markets, transport costs
and the contractual conditions affecting returns on investment. After
a discovery has been made, the producer must consider the size of
the field and the productivity per well.

The institutional framework within which a developer operates
affects the economic viability of an oil development project. This
includes the laws, policies and regulations that affect the conducting
of exploration activities, the process of production itself, as well as

transportation or movement of the output within or away from the country. This would include policies on conservation and the environment. It also includes such policies, laws and regulations that affect the incoming and outgoing movement of capital funds and, when not specified in the exploration or development contract, the level of net income after taxation.

The economic viability of developing and producing an off-shore field must, therefore, take into consideration many factors. Basically these factors are the expected returns from the field (and therefore the size of the reservoir) and the desired return on investment, but to compute the latter other factors like 'government take' (in taxes, royalties, bonuses, production shares, etc.), climate and drilling depth, figure prominently in decision-making. The relationship between these factors, in addition to the climate and depth elements, are shown in Table 6.1. Anything that affects net profitability affects the economic viability of a given reservoir. For example, as 'government take' increases, the economics of various prospects deteriorate markedly. Also, in this example, the return on investment assumed for the calculations is 20 per cent. If the return desired is reduced to 15 per cent, the rating of the economic viability of certain reservoirs considered uneconomic in the previous calculations would improve.

3. *Risk factors.*[5] In an area where exploration and development have already established the presence of a number of petroleum fields, sufficient insight into the geological setting for oil accumulation may have been obtained to allow predictions of the possibilities of other adjacent areas. Thus, it may be possible to extrapolate statistically from the area where reserves have been 'proven' to estimate the probabilities of expected approximate sizes of accumulations that could be present. In this case, the 'success ratios' (or the ratio of successful wells to total exploration wells drilled) in the known areas may be used to project the chance of discovery in adjacent areas. By taking into consideration regional geological conditions, the probability of the occurrence of accumulations is taken into account, as is the probability that such accumulations, if present, can be discovered.

However, in an area lying at a distance from a proved area, and on

TABLE 6.1

INDICATED ECONOMICS OF OFF-SHORE EXPLOITATION UNDER ALTERNATIVE LEVELS OF GOVERNMENT FINANCIAL TAKE

Water depth (metres)	Large reservoir Climatic conditions			Medium reservoir Climatic conditions			Small reservoir Climatic conditions		
	Mild	Moderate	Severe	Mild	Moderate	Severe	Mild	Moderate	Severe
*20% Return on Investment, Low Government Take**									
200	E	E	E	E	E	E	E	E	-1.5E
500	E	E	E	E	E	-1.5E	E	-1.5E	-2.5E
1 000	E	E	-1.5E	-1.5E	-1.5E	-2.5E	-2E	-2.5E	-4.5E
20% Return on Investment, Medium Government Take+									
200	E	E	E	E	E	-1.5E	E	E	-2E
500	E	E	-1.5E	E	-1.5E	-2.5E	-2E	-2E	-4E
1 000	E	-1.5E	-2E	-1.5E	-2E	-3.5E	-3E	-3.5E	-6.5E
20% Return on Investment, High Government Take≠									
200	E	E	-1.5E	E	E	-1.5E	E	-1.5E	-2.5E
500	E	-1.5E	-2E	-1.5E	-1.5E	-2.5E	-2E	-2.5E	-4.5E
1 000	-1.5E	-1.5E	-2.5E	-2E	-2E	-4E	-3E	-3.5E	-6.5E

Notes: (1) E = Economic (20% ROI as a guide) at projected long-term value of seabed crude oil (U.S.$11-$13/Bbl. in constant 1974 dollars).

(2) Negative multiples of E (example, -2E) are uneconomic and indicate the degree by which such cases would fail to meet assumed economic standards.

* No royalty or bonus and tax provisions similar to those that currently apply to U.S. federal off-shore leases.

+ Substantial royalty, no bonus and moderate taxes.

≠ Substantial royalty, moderate taxes and U.S. $1.00/Bbl. additional government payment.

Source: National Petroleum Council, *Ocean Petroleum Resources*, 1975, Table 7, p. 28.

which minimal geological and geophysical exploration has been done, the employment of statistical extrapolations are no longer useful, because every geological province has at least a somewhat different geologic history. Thus, the probability that each of the geological factors favouring oil accumulation is present cannot be properly estimated—and therefore neither can one estimate the probability of success.

The exploration success ratio in the off-shore areas of Indonesia between January 1973 and May 1974 was reported in the CCOP's eleventh session to be 1:4, which is high compared with about 1:10 for both on-shore and off-shore exploration in the United States.[6] Malaysian off-shore areas were reported to have success ratios that were as good as those in Indonesian waters.[7] There is a tendency to flaunt such success ratios as indications of the probability of individual success. The success ratio is, however, only useful for assessing the overall potential of a specific area that has been fairly developed or tested. In addition, statistical success ratios apply to a set of samples, so that at an industry success rate of 10 per cent for every 50 wells drilled, an individual company still stands a 12 per cent chance of drilling 20 dry holes in succession.[8] It is thus possible for an explorationist to make a large profit on his investments in a short time, or to lose his investments totally, without hope of any returns.

The technology for detecting accumulations of oil and gas has, of course, continually been improving. One of the more recent breakthroughs was reported in the magazine *Science* in 1974.[9] The 'bright spot' technique which was reported to have been extensively used offshore in the Gulf of Mexico, Nigeria, and Indonesia, was said to have improved success rates. At the time of reporting (1974), 60 to 80 per cent of the wells drilled using this technique were reported to have been successful. A significant feature of 'bright spot' analysis was said to be the potential for finding hydrocarbons in regions where there are none of the structural features commonly associated with petroleum, such as salt domes and anticlines. The technique allows the detection of oil and gas in stratigraphic traps.[10] This technique has, however, been found not to be universally useful and foolproof,[11] so that the risks of failure as well as success in locating a reservoir remain almost unchanged. There still remains no better alternative to drilling into a trap to find out whether oil or gas exists, and in what amounts.

COSTS OF OFF-SHORE EXPLORATION AND DEVELOPMENT

Assuming that the above conditions all favour exploratory drilling, test wells are then drilled — three or four or more. If the initial exploratory wells are dry, additional study is engaged in before further investment in the area is made. If the exploratory wells yield oil or gas, a chain of new investment begins. Development wells are drilled to delineate the size of the discovery. If the reservoir of oil or gas is large enough to justify the capital investment related to production, all related producing investments are made.

Investment costs required to bring about production of petroleum resources may be grouped for convenience into two classes: exploration and development costs.[12]

Exploration costs (also sometimes referred to as the cost of locating new reserves) include those elements involved in determining the location of hydrocarbons prior to drilling development wells and initiating production. As stated earlier, exploration activities generally begin with geophysical and geological surveys, and conclude with the drilling of exploratory wells. Geological and geophysical surveys are sunk costs in terms of investment decision, that is, these costs are not capitalized but are deducted from income in the year of expenditure.[13] In the second stage of exploration, the cost of exploration is a function of the cost of each exploratory well and the number of wells which are drilled on any given structure or tract. The number of wells required to explore a structure varies significantly between structures.[14]

Development costs involve installing production wells and all facilities related to initiating production activity, transporting output to shore facilities, and abandoning a depleted field. They are a function of platform costs, water depth, drilling depth, dry hole risk factors, drilling difficulty, completion costs, labour costs, climate and others.

Exploration and development of a field in off-shore tracts involve some methods and equipment different from those used on land areas. Exploration on the outer continental shelf includes both geophysical exploration and exploratory drilling to gather information about the geology of the sea-bottom, in order to assess the petroleum potential of the area.[15] Geophysical exploration includes passive reconnaissance techniques, such as air and ship-borne magnetic and gravity surveys or similar measurements of hydrocarbon seepage into the atmosphere. It also includes active surveying techniques such as seismic analysis, bottom sampling and bottom coring.[16] Exploratory drilling into off-shore geologic formations involves mounting drilling equipment on a platform (that is, on a barge, a drillship, a semi-submersible platform, or a jackup platform).[17] Drilling methods and most of the drilling equipment are identical to those used on-shore.

Cost figures are not readily available to allow comparison of exploratory geological/geophysical work on land versus similar studies of the seabed. To get an idea of geophysical costs, a small Philippine company and its foreign partners spent US$1 million over a period of five years (1970-1975) for aeromagnetic data acquisition and interpretation and marine seismic surveys on concessions off the coasts of Palawan.[18] In the Far East, expenditures for geological and geophysical surveys and lease rentals were estimated to be US$225 million in 1973 for both on-shore and off-shore exploration.[19] At that point off-shore exploration had reached a peak in South-East Asia and accounted for a large portion of these expenditures. This figure was almost twice the amount spent in 1969 (US$125 million) and three times the amount spent in 1964 (US$75 million), when off-shore activity was just starting in this region.

In Chapter II the contractual cost of drilling an exploratory well off the coast of Burma in early 1976 was put at US$40 000 per day and about US$5 million on the average. The U.S. National Petroleum Council also presented estimates of the cost of exploratory drilling to 200-metre depths in moderate climatic conditions (defined as similar to those in the Gulf of Mexico and the South Pacific, areas which include monsoons). Assuming a rig capital cost of US$20 million, these costs expressed in 1974 constant dollars are shown in Table 6.2. These costs are exclusive of geological and geophysical survey costs that precede drilling, costs related to acquisition of the rights to conduct exploratory drilling, or capital costs.

Assuming the same well depths and geological conditions, explo-

TABLE 6.2
BASE CASE EXPLORATORY DRILLING EXPENDITURES
(Thousands of 1974 Constant U.S. Dollars)

Item	Amount
Drilling expenditures — day rate of $27 M/D x 80 drill days (10-12 000 foot well)[1]	US$2 160
Equipping expenditures — day rate of $27 M/D x 7 equipping days	189
Tubular goods	264
Wellhead	50
Testing	26
Other	25
Total per well drilling and equipment expenditures	US$2 714

Note: The Base Case is for 200 metres water depth, moderate climate, expressed in thousands of 1974 constant dollars.

[1] The day rate is directly related to the cost of the rig and is intended to cover depreciation, insurance, interest expense, variable general and administrative expense, direct operating expense and a financial return to the rig owner. A rig capital cost of US$20 million is assumed.

Source: Table 5 in National Petroleum Council, *Ocean Petroleum Resources* (Washington, D.C., 1975), p. 24.

ratory drilling expenditures differ according to the location of the site, including water depths and climate. Table 6.3 shows the effects of water depth and severity of climate on exploratory drilling costs. Realistically, perhaps the index for the 4 000-metre depth is meaningless for the foreseeable future. In South-East Asia, the indexes related to moderate climate conditions are relevant, that is, conditions similar to the Gulf of Mexico *et al.* group. This will include areas south of Java and north of Brunei, or the central Philippines, which are subject to typhoons similar to the Gulf of Mexico hurricanes.[20] Adjustments upward must, however, be made for differences in mobilization logistics in South-East Asia. Costs are higher when a rig

TABLE 6.3

OFF-SHORE EXPLORATION DRILLING EXPENDITURE INDEX
(1.0 = U.S. $2.7 Million per Well in 1974 Dollars)

Water depth (Metres)	Climatic conditions				
	Mild (1)	Moderate (2)	Severe (3)	Ice laden	
				75% (4)	100% (5)
200	0.8	1.0	1.8	2.3	4.6
500	1.0	1.3	2.1	2.8	5.4
1 000	2.5	2.8	3.6	4.3	6.4
4 000	3.8	4.0	4.3	5.6	7.5

Notes: Typical of the various climatic conditions are:

(1) Senegal, Gabon, Honduras, Mediterranean, Java Sea, Persian Gulf.

(2) Gulf of Mexico, South Atlantic, South Pacific,* North-west Australia,* Sea of Japan,* Yellow Sea.

(3) North Sea, Bay of Biscay, South Australia, Gulf of Alaska,* North Atlantic, North Pacific, West Coast of Canada, Nova Scotia. +

(4) Bristol Bay, Alaska,* West Greenland. +

(5) Arctic Ocean, + Chukchi Sea. +

* Earthquakes

+ Icebergs

Source: National Petroleum Council, Ocean Petroleum Resources, 1975, Table 6, p. 26.

has to be moved from one area to another and when waiting, and higher start-up costs are involved. In this region, exploration activity is not concentrated in one area but may involve transfers from, for example, Balikpapan to the Gulf of Martaban, or even from the Red Sea back to the South China Sea.[21]

In contrast to exploration activity, which usually involves only a few wells, development of a field that is considered commercially viable means drilling a large number of wells.[22] In addition, associated facilities must be installed for gathering the oil, separating unwanted components from the crude oil, storage, transportation, as well as facilities for safety and environmental protection. Thus, in off-shore production, a typical unit is viewed as two multi-well platforms and

TABLE 6.4
DEVELOPMENT AND PRODUCTION ESTIMATED
EXPENDITURE REQUIREMENTS
(1.0 = U.S. $95 Million per System in 1974 Dollars)

Water depth (Metres)	Climatic conditions				
				Ice laden	
	Mild (1)	Moderate (2)	Severe (3)	75% (4)	100% (5)
200	0.9	1.0	2.8)	Unknown but	
300	—	—	6.2)	estimated to be	
500	2.7	3.0	—)	substantially	
1 000	4.3	4.8	10 2)	greater than severe.	

Note: Typical of the various climatic conditions are:
  (1) Senegal, Gabon, Honduras, Mediterranean, Java Sea, Persian Gulf.
  (2) Gulf of Mexico, South Atlantic, South Pacific,* North-West Australia,* Sea of Japan,* Yellow Sea.
  (3) North Sea, Bay of Biscay, South Australia, Gulf of Alaska,* North Atlantic, North Pacific, West Coast of Canada, Nova Scotia. +
  (4) Bristol Bay, Alaska,* West Greenland. +
  (5) Arctic Ocean, + Chukchi Sea. +

  * Earthquakes
  + Icebergs

Source: National Petroleum Council, *Ocean Petroleum Resources*, 1975, Table 8, p. 30.

associated facilities. Table 6.4 gives an index of expenditures for development and production at different water depths and different climatic conditions, for a base system of US$95 million (1974 constant dollars). A recent figure indicated that a platform large enough to take care of about twenty production wells over a 9 square mile (23 square kilometres) tract in an open sea would cost about US$25-40 million.[23] The production wells themselves are somewhat less expensive than the platform. In 1976, the cost of a fully equipped offshore well in the United States was estimated to range from less than US$1 million to over US$3 million, depending on productive capacity, field size, and various geological circumstances.[24]

TABLE 6.5

COMPARISON OF ON-SHORE AND OFF-SHORE COSTS AT
VARIOUS DEPTH INTERVALS, 1973

Well depth intervals (feet)	Louisiana		Texas	
	On-shore	Off-shore	On-shore	Off-shore
0-1 249				
Average footage	790	1 180	—	—
Cost $	8 384	231 044	—	—
3 750-4 999				
Average footage	4 313-(low) 4 363 (high)	4 478	4 217- 4 323	4 705
Cost $	27 496- 98 371	435 707	43 927- 63 945	320 998
17 500-19 999				
Average footage	18 567	17 576	—	—
Cost $	1 414 936	1 466 236	—	—

Source: American Petroleum Institute *et al.*, *Joint Association Survey of the U.S. Oil and Gas Producing Industry* (Washington, D.C., February 1975).

The costs of drilling and equipping a producing well off-shore are generally higher than on-shore. To get an idea of the magnitude of the difference, a comparison is made in Table 6.5 of actual costs in 1973 for on-shore and off-shore wells (including platform costs) at given depths in the two largest producer states in the United States. The high initial capital expenditure related to off-shore production is clearly seen in the relationship between average costs and depth intervals. In this example, the differential between off-shore and on-shore costs narrows abruptly with depth. This is because the producing wells in Louisiana and Texas were at that time in relatively shallow waters.[25]

The magnitude of capital investments required as early as the exploratory stage but even more so in the development stage has required the influx into South-East Asia of foreign capital to support such ventures prior to the revenue-earning stage. Thus, in the Philippines, the earnestness of the government in its search for oil has also implied seeking the participation of foreign firms as partners to local companies. These foreign firms would conduct the exploration and, hopefully, would finance the infrastructure associated with the production stage. Burma has also opened its oil exploration activity in the off-shore areas to foreign companies, as have Thailand and the Indo-chinese countries.

In 1976 the World Bank made estimates of the capital investments that would be required in non-OPEC developing and oil-producing countries to meet increased oil production capacity. Using their results and pro-rating Burma's oil reserves as of 1974 against the projected capital investment requirements of low-income developing countries (which includes Burma, India, and Pakistan),[26] Burma would require an average capital investment of US$44 million annually for crude production facilities alone, and an additional US$1 million for pipelines to transport such production from currently known reserves.

In Malaysia, Exxon is said to have spent US$250 million in exploration work and production capacity by the time it started to lift oil and receive some returns on its investment.[27] Exxon's first crude oil from Malaysia came from the Tembungo field in Sabah in late 1974,[28] although this company actively entered the Malaysian exploration picture in the early 1960s[29] — a long wait at first glance, but a welcome reward in the light of dry holes experienced elsewhere in South-East Asia (such as in off-shore Burma and the Philippines).

Some Economic Impacts of Petroleum Resource Development

The social costs of conflicts over property rights will be better appreciated if the types of economic variables that would be affected by the development (or non-development) of South-East Asia's petroleum resources are understood. The impacts will be regional as well as national. Indicators may be formulated and disaggregated in various ways, but they may be reduced to a few basic types:[30]

(1) the petroleum resource availability effects, that is, the degree of dependency on foreign energy resources over time (and

therefore the effect on energy-intensive development programmes);

(2) the foreign exchange effects;

(3) governmental revenue effects;

(4) the direct domestic input effects, that is, employment and income in energy-related sectors;

(5) the indirect factor mobilization inducement and productivity
effects, including employment and income in sectors which
may be adversely affected by petroleum resource development;

(6) allocative effects on economic resources;

(7) income distribution effects.

The first type has already been discussed in the chapters on the demand and supply of petroleum.

The impact that development of off-shore oil resources will have
upon each of the foregoing economic variables will depend to a large
extent on the pertinent national policies adopted, as such policies
would influence, first, the rate of investment in exploration and development, and second, the rate and timing of production, production costs, and the degree of resource recovery.

FOREIGN EXCHANGE EFFECTS

The foreign exchange effects of domestic oil production may be
viewed in terms of foreign exchange earnings for oil-surplus nations;
or foreign exchange savings for smaller producers whose production
is less than or equal to domestic consumption. Under the first category are, at present, Indonesia and Brunei; in the future Malaysia
is also expected to be a net exporter. The contribution of crude oil
and petroleum product exports to foreign exchange reserves in these
countries was mentioned earlier. Such earnings for the years 1970,
1973, and 1975 are also shown in Table 6.6. No official data were
available, at the time of writing, on the export earnings of Brunei
in 1975, but unofficial reports indicated that in 1976 such earnings
would amount to about B$1.6 billion or US$600 million.[31] It will
be noted in Table 6.6 that Malaysia, whose off-shore fields came
onstream only in recent years, earned just under US$4 million in
1975. This amount accounted for over 8 per cent of its total foreign
exchange earnings. As noted earlier, Burma's production is currently

TABLE 6.6

PETROLEUM EXPORT EARNINGS: BRUNEI, INDONESIA, MALAYSIA
(SELECTED YEARS)
(Million U.S. Dollars)

Country	1970			1973			1975		
	Petroleum exports (X_O)	Total exports (X_t)	X_O as % of X_t	Petroleum exports (X_O)	Total exports (X_t)	X_O as % of X_t	Petroleum exports (X_O)	Total exports (X_t)	X_O as % of X_t
BRUNEI[a]	90.4	94.2	96.0	336.5	337.7	99.6	n.a.	n.a.	—
INDONESIA	446.3	1 108.1	40.3	1 608.7	3 210.8	50.1	5 282.1	7 103.2	74.4
MALAYSIA[b]	66.0	1 682.3	3.9	110.1	2 958.7	3.7	355.2	4 242.6	8.4

Sources: Indonesia and Malaysia: International Monetary Fund, *International Finance Statistics* (Washington, D.C.), July 1976.
Brunei: Brunei, Economic Planning Unit, *Statistics of External Trade*, 1973.

[a] Conversion rates: 1970 — B$3.1 = US$1; 1973 — B$2.46 = US$1.
[b] Conversion rates: 1970 — M$3.0612 = US$1; 1973 — M$2.4433 = US$1; 1975 — M$2.4018 = US$1.

meeting most of its domestic needs; some observers predict that Burma will be a net exporter in the 1980s.

Other countries are, however, not so fortunate. Thailand's production is minor compared to its requirements. All other South-East Asian countries spend a considerable portion of their foreign exchange earnings on crude oil imports. In fact, for oil-importing developing countries in general, most of which have been experiencing international reserve shortages, the import costs of crude oil have had economically disruptive allocation effects. The burden of oil imports on all non-producing developing countries rose from less than 10 per cent of export earnings in 1973 to about 17-20 per cent of export earnings in 1974 and 1975. In absolute terms, oil import costs of these countries rose from US$4 billion in 1973 to US$16 billion in 1974 and US$17 billion in 1975.[32]

Table 6.7 shows changes in the oil import costs of two South-East Asian countries—Thailand and the Philippines—in relation to their export earnings for the years 1965, 1972, and 1974. It will be noted that the impacts of quadrupled oil prices have been more severe than those suffered by all non-oil producing LDCs aggregated together. The cost of oil imports in 1974 took up more than double its share of export earnings in 1972 in the Philippines, and just about double in Thailand, despite the more than 100 per cent increase in total export earnings of both countries. Under those circumstances, the total social costs of not developing one's own resources may be greater than the actual costs of such development. Quantification of the loss of welfare is as usual difficult, but there are, no doubt, real losses. The increase in oil prices has the following effects: if foreign exchange reserves are kept at the normal level without additional borrowing, it means having less resources left after paying for oil imports to buy other goods. If a developing country decides to keep its ability to buy other goods at the same level, it means having to borrow additional foreign exchange to pay for the higher cost of oil imports. In either case, there is an additional burden that a country may not be willing to bear indefinitely. In fact, from the time oil prices began to rise sharply in the 1970s, oil-importing countries could foresee that their balances on current account would deteriorate significantly. A number of them decided to augment their offi-

TABLE 6.7

IMPORT COSTS OF OIL VS. TOTAL EXPORT EARNINGS: PHILIPPINES AND THAILAND (SELECTED YEARS)

(Thousand U.S. Dollars)

Country	1965			1972			1974		
	Oil imports (Mo)	Total exports (X)	Mo as % of X	Oil imports (Mo)	Total exports (X)	Mo as % of X	Oil imports (Mo)	Total exports (X)	Mo as % of X
PHILIPPINES	68 349	768 400	8.9	137 750	1 145 000	12.0	811 000	2 725 000	29.8
THAILAND	63 671	593 221	10.7	149 759	1 081 298	13.8	598 714	2 398 238	25.0

Source: U.N., Proceedings of Symposium of Petroleum Resources in Asia and the Far East, Vol. III, pp. 43, 44 for 1965 import data.
P.D. Gaffney et al., 'Economic Appraisal of the Potential Petroleum Resources of the Asian Pacific Region', February 1976, for Philippine 1974 consumption.
Bank of Thailand, Monthly Bulletin, June 1975, Table III-6, for Thailand 1972 and 1974 import and export data.
Central Bank of the Philippines, Annual Report, 1975, for Philippine export data.
U.N., Statistical Yearbook for Asia and the Far East, 1973, for other import and export data.

cial monetary reserves by recourse to borrowing abroad in order to finance anticipated deficits.[33] This route had to be taken because the bulk of imports of developing countries are essential items — food and capital goods for development projects.

There has been sufficient concern in international bodies as well as in academic circles about the effects of increased oil prices on the economic growth of both developed and developing countries, and the foreign exchange requirements of oil-importing developing nations. In 1973, a World Bank paper on this subject estimated the oil import bills of countries like the Philippines and Thailand to rise by 1980 to US$860 million and US$620 million, respectively, by their low estimate,[34] numbers which already seemed onerous. The actual 1974 costs of these two countries closely approximated these 1980 projections.[35] Numerous studies have since been made with the same purpose in mind.[36] They all point to the fact that imported oil is becoming increasingly burdensome, in terms of what each country can afford.

The development of indigenous resources, regardless of relative costs, to prevent a drain on foreign exchange is one that economists would consider inefficient when considered by itself, if the cost of developing one's resources exceeds that elsewhere. According to the theory of comparative advantage, a country should produce that good that it can produce cheaply and trade this for goods that it is less efficient in producing. However, this presumes, among other things, a perfectly competitive situation where the prices of goods reflect their marginal costs. The international oil market is far from competitive, however, and world oil prices are far in excess of the direct marginal costs of production. The high oil prices have constituted mere transfers of wealth without compensation, from some developing countries that are less well-endowed to developing countries endowed with oil resources.[37] The direct production cost is, therefore, irrelevant; the relevant cost to a user country is the cost of imports versus domestic production cost.[38]

But even in the case where production cost may be higher than import cost, the societal cost of importing oil may justify domestic production if there is a divergence between social and private costs. The acquisition of foreign exchange reserves, after all, is not costless.

Developing countries normally have inadequate foreign exchange reserves to finance their development programme, because their export earnings fall short of required import expenditures. As a result they have to borrow money. Such external debt, it has been observed, is serviced with 20 to 25 per cent of exports for periods lasting twenty years or more.[39]

The exchange reserves benefits from developing one's indigenous reserves extend beyond the reduction of the loss of foreign exchange reserves. The development of resources, in fact, could mean an inflow of foreign exchange in the form of capital funds for the development of the oil reserves themselves and other foreign exchange inflows for the ancillary activities related to the development of such resources.

DOMESTIC GOVERNMENT REVENUES FROM PETROLEUM DEVELOPMENT

The domestic revenues accruing to a government as a result of petroleum resource development may result, directly, from taxation of producing and marketing operations or, indirectly, from taxation of incomes earned by individuals or firms employed in or servicing the petroleum industry. The importance of such revenues in terms of total fiscal earnings will be dependent on the importance of the industry to the economy. In Brunei, for example, where 1976 income from the petroleum industry was expected to be US$545 million,[40] the present affluence of the state is, without question, the result of offshore oil and gas discovery and production. But this also is so because of the size of the economy relative to its petroleum resources. Seen without reference to its land-size and population, its production and reserves are not so large as to evoke awe. But for a state with a total area of only 2 226 square miles (5 765 square kilometres), and a population of only 150 000 in 1975,[41] oil and gas revenues from the oil industry have allowed, for the present and foreseeable future, a welfare sultanate that provides free education and medical care, and that pays pensions to widows, the aged, etc., without the burden of income taxation on its nationals.[42]

In Indonesia also, the taxation of the oil industry is an increasingly important source of government revenues (see Table 6.8). In the

TABLE 6.8

FISCAL REVENUE FROM OIL IN INDONESIA
(SELECTED YEARS)
(Million U.S. Dollars)

Item	1966	1972/3	1974/5	1975/6	1976/7 (projected)
Corporate tax on oil companies	(2	477	2 345	3 009	3 992
Oil product receipts	(	76	-38	0	0
Non-oil revenue	34	857	1 900	2 393	2 763
Total non-aid revenue	36	1 410	4 207	5 402	6 755
Oil corporate tax as % of total	less than 5	34	56	56	59

Source: U.S. Embassy, Economic Section, *Indonesia Petroleum Report* (Jakarta, Indonesia), issues dated June 1975 and 21 June 1976, Table 3.

past two years, such taxes (exclusive of share in kind) from producing and marketing operations have provided 56 per cent of fiscal earnings, compared to less than 5 per cent ten years ago. In the fiscal year 1976-7, this share was projected to increase to close to two-thirds of total fiscal earnings. This contribution from the oil sector is especially remarkable in view of the relatively minor share of oil production in the Gross Domestic Product (GDP). Since 1973, the mining sector (which includes both oil and non-oil outputs) has contributed only around 12 per cent of the total GDP.[43] As a revenue source, the oil sector in Indonesia has been a major vehicle for financing the country's economic development programme, in terms of both domestic fiscal earnings and foreign exchange earnings. Such revenues are a major factor in plans to improve the country's productive capacity and to finance its social programmes.[44]

FACTOR INPUTS AND MOBILITY EFFECTS

Conceptually the major categories of factor inputs in the production function are land, capital, human resources, and technology. In the practical sense, however, technology may be subsumed in both capital investments and human resources. The impacts of petroleum development on capital and human resources would be direct and indirect.

The direct impact of off-shore development on economic variables depends on the level of activity or the size of the operation it generates. This in turn depends mainly on the location of the oil/gas fields; expected size of the reservoirs, estimated production rates, and type of production; and whether oil and gas are likely to be transported ashore by pipeline or tanker. The magnitude of the effect on local employment and income depends on the foregoing factors, plus the following: whether rigs and platforms will be constructed locally or imported; and whether only necessary or optional development facilities associated with off-shore petroleum production are planned. Necessary facilities include treatment plants and separation facilities; optional facilities include petrochemical plants.

Direct impacts on capital and labour resources. The petroleum industry is capital-intensive, and the direct effect of the development of a nation's resources would be to employ and reward capital resources heavily relative to its human resources. In addition, in South-East Asia, capital funds for exploration and development must, in the early stages, come from foreign sources. The same is true for skilled human resources.

Table 6.9 shows annual expenditure by foreign oil contractors on exploration and development of productive capacity in Indonesia for the years 1969 to 1976. It will be noted that by 1975 the annual level had reached US$1 billion. The cumulative total over the eight-year period was estimated to be close to US$4 billion.[45] At the end of 1973 about 46 per cent of these expenditures were for oil-search activities. (This proportion dropped considerably in 1974 to 33 per cent, however, as less attention was spent on exploration and more on the development of discovered fields.)[46]

Being capital-intensive, the industry's high levels of expenditures had only a minor direct impact on employment. Arief shows that in 1971 the mining industry employed only 0.2 per cent of Indonesia's total labour force.[47] At the time of writing (presumably 1975), he estimated that only slightly over 58 000 workers were employed in the oil sector,[48] and he proposed a capital/labour ratio of US$33 000:1

TABLE 6.9

EXPLORATION AND DEVELOPMENT EXPENDITURES BY
FOREIGN OIL CONTRACTORS IN INDONESIA, 1969-1976
(Million U.S. Dollars)

Year	Amount[a]	Year	Amount[a]
1969	78	1973	393
1970	139	1974	807
1971	232	1975	1 047
1972	286	1976[b]	1 000

Source: U.S. Embassy, Economic Section, *Indonesia Petroleum Report*
(Jakarta, Indonesia), issues dated June 1975 and 1976, Table 4.

[a] Does not include investment costs of two large liquefied natural gas projects or other capital projects.

[b] Estimated, and does not include outlays for field development of two
LNG projects.

as the capital intensity of that sector.[49] This was about five times the
capital intensity of foreign investment projects in general, and about
nine times that of domestic investment projects.[50]

The above discussion refers, of course, to activity related to both
on-shore and off-shore petroleum activity in Indonesia.

Related developments and impacts. Production implies marketing of
the output. Therefore, transportation and storage facilities have to
be provided, once oil or gas starts flowing. The petroleum has to
be processed, and then transported to the shore by pipeline or by
tanker. Gas is always moved by pipeline but oil may be transferred
by either tanker or pipeline depending on the distance from the
shore. If tankers are used for the oil produced, deep moorings and
oil storage facilities have to be provided.

Assuming considerable discoveries in extensive fields, oil companies
may need one or more pipelines to pump the oil ashore into storage
tanks. From the storage tanks crude oil can be pumped through
pipelines to existing or newly built refineries, or shipped to other
refineries. Oil companies are less likely to use pipelines for small

discoveries in relatively small stratigraphic structures. Instead oil may be stored beneath the production platform, where tankers collect the crude oil and transport it ashore for further processing. All this activity requires on-shore support to service the drillship and production operations. Shipyards, machine shops, drilling and non-drilling materials and supplies, and other phases of the service sector normal to any economic activity are needed to support production.

As already noted, the oil mining and refining industry is highly capital-intensive. Hence, in off-shore oil and gas development the number of people directly employed is not very large. The indirect employment impacts may, however, be considerable. In the absence of adequate data on the South-East Asian experience, it is possible to obtain an idea of the effects of off-shore development on the economies of coastal states by studying the changes that have occurred in other off-shore areas.

The following data from the state of Louisiana in the United States gives an idea of the relationship between direct and indirect employment impacts of off-shore production.[51] Off-shore Louisiana produced the equivalent of 1 773 000 barrels per day of oil and gas in 1972; the number of people directly and indirectly employed was as follows:

Persons directly employed in off-shore production	8 000
Persons directly employed in oil industry related areas	30 000
Persons indirectly employed	76 000

Just as there are indirect employment effects generated by investment in petroleum resource development, there are also indirect investment effects. The level of impacts again will depend on some variables: the size of the operation itself which will determine the size of the supporting service industries and the size of the increase in other related manufacturing activities. These in turn will depend on the existing infrastructure and industrial capacity in the area. Again using the example of the state of Louisiana for the effect of off-shore production on the location of investments, between 1938 and 1971, 80 per cent of all new investment in manufacturing in

Louisiana was made in the coastal zone, and these manufacturing investments were oil-related.[52]

Because of geographical, economic, social, and political conditions, such indirect impacts of off-shore oil development may not always be contained within the country developing its resource. This has been true in South-East Asia in the past ten years. Singapore has provided the base for servicing not only oil exploration but also development activities in the region. Thus the indirect economic impacts of petroleum development have been experienced by Singapore to a major extent. A boom brought an influx of expatriate personnel from supply and service companies that required both physical facilities (office and residential) and labour resources. A retrenchment of activities in South-East Asia since mid-1974, following uncertainty in future exploration and development plans of oil companies, was accompanied by departures of expatriate personnel and a slack in exploration-related activities in Singapore.

The use of Singapore as a base for such activities may be considered as a special case. As fields which have been discovered are developed, certain supply and service bases may choose to be nationally located for various reasons, primarily economic (short-term or long-term). Thus some service and supply companies have been relocating to Malaysia in preparation for development of its off-shore fields in the South China Sea; as this is written, a port and off-shore supply base is under construction in Kuantan. The development of Batam Island should also concentrate some of the support facilities for Indonesian oil development on national shores, and thus allow the country to enjoy the spin-offs that result from petroleum development activities.

ALLOCATIVE EFFECTS

The development of the petroleum resources of a nation requires the allocation of some of the economy's resources to such activity. Capital and human resources must be employed. If there is a surplus of either, this poses no problem for the achievement of national goals. But where no surplus exists, it means the allocation of resources from some project or programme to this sector. In developing countries, manpower resources are generally in over-supply. Although

there may be an initial shortage of skilled manpower for the petroleum sector, the possibility of training and education to meet such
needs presents a minor problem over the longer term. Capital or
investment resources, however, are generally in short supply in developing countries relative to capital needs for development, and
South-East Asian countries are no exception. Thus development of
indigenous petroleum resources to meet energy needs requires allocation of scarce investment resources away from other projects.

The question of allocating such scarce resources to energy resource development without sacrificing other economic development
programmes is a genuine concern. A study at the World Bank
addressed the question for all non-OPEC developing countries vis-à-
vis all forms of primary energy production.[53] That study pointed out
that, 'should the minimum required amounts of capital not be met
sufficiently or in time, the non-OPEC developing countries will have
to face steeper reductions in economic growth or further reductions
in use of energy per unit of product, which entail endangering their
longer-term prospects for faster growth and industrialization'.[54]

The expenditures required to bring about production of off-shore
petroleum have been discussed above. There are also other costs
that must be met from the nation's resources. Some of these are the
costs of planning for the regulation of on-shore and near-shore
petroleum-related facilities; the costs of coastal zone management;
and the costs of public facilities to ameliorate the adverse environmental impact of petroleum-related activities. Some of these costs
may be incurred by the foreign investor, where such requirements
are imposed by the national government as one of the contractual
conditions for resource development. Such facilities may also generate tax revenues that might well cover these costs. There is, nevertheless, the timing question, that is, capital resources must be spent
initially to take care of these facilities over a short period of time
when other economic and social priorities require attention.

INCOME DISTRIBUTION EFFECTS

Any economic activity that gives rise to earnings results in some
redistribution of income. How it is redistributed and what criteria
are used for judging the desirability of such redistribution in terms of

the ethics of the relevant society will depend on specific activities and specific social norms.

Development of off-shore petroleum resources in South-East Asia suggests that the redistribution will not take place in favour of private property owners of the acreage under which petroleum is found and produced, as is the case in on-shore areas of countries where such resources are not state-owned. The producer's surplus or economic rent associated with natural resource exploitation would, in this case, accrue to the state government.

Still, there would be a redistribution of income resulting from the investment of huge funds for the development of petroleum fields, and in the growth of economic activity in the coastal area. The significance of the shift would, of course, depend on growth in the rest of the economy. There would be returns to capital and human resources employed in off-shore development directly and to the service and manufacturing industries attracted to the coastal zone. It is also conceivable that this zone would grow faster than some other industrial or rural areas in the country.[55] There may be a redistribution in favour of the highly skilled, the category that would be employed in the petroleum industry. How the results in this aspect will be judged will depend on the objective function of the state in relation to its decision to develop its petroleum resources. What are its criteria? In the formal sense, the general rule might be an increase in social marginal productivity. This criterion must, however, be interpreted within the total dynamic complex of development. Is it higher national income or per capita income that is to be maximized? Does this maximization involve a more unequal distribution of income? For whom?

As stated, the petroleum industry is a high-technology industry. One of the causes of income inequality has been shown to be the degree of dualism in the structure of many developing countries.[56] Where a technologically and institutionally backward sector and a technologically advanced and well-organized modern sector exist side by side, the inequality was found to be greater where the disparity between these two sectors was more pronounced. The author, in a less intensive study of the relationship of Singapore's petroleum refining sector to the rest of the manufacturing industry, found a petroleum/total industry ratio of average employees' remuneration that averaged 4.5 between 1970 and 1972 and 3.9 between 1973 and 1975.[57]

Since the development of a modern, technologically advanced sector is an essential ingredient of economic growth, then it may be argued that it is not the growth of the technologically advanced sector as such that causes inequality in the distribution of income, but economic growth itself. This dependence between economic growth and inequality of income is not arguable, however, since growth results from increased productivity and is the fruit of differential rewards for skill, entrepreneurial abilities, and initiative, among other things. A redistribution of income to solve the inequity problem may be sought in other ways, the discussion of which is beyond the scope of this topic.

Some Environmental Impacts of Off-shore Petroleum Development

Petroleum exploration and production impact on the environment. To a large extent, the environmental costs of off-shore development are a function of the level and nature of production activity, and the location of such activity in relation to marine, bird, and other wildlife. Some details relating to environmental damage while exploring for oil in South-East Asia were touched on earlier, and general impacts are discussed in this section.

During hearings in the United States concerning the development of the Atlantic outer continental shelf, witnesses testified that the risks to the marine environment of off-shore petroleum development may be minimized. One witness stated that in the North Sea the best technology permitted development, production and transportation of petroleum at 'acceptable' levels of risk.[58] Other witnesses pointed to the U.S. experience where in drilling 18 000 off-shore wells, only four major spills had been experienced (the definition of what is 'major' is subjective).[59]

The sources and extent of ocean petroleum pollutants are given in Table 6.10. It shows that off-shore production accounted for only about 1 per cent of the total sources of petroleum pollutants in the

TABLE 6.10
SOURCES OF OCEAN PETROLEUM POLLUTANTS

Source	Petroleum pollutants (MB/D)	Per cent of total
Land-based activities	67	54
All vessel activities	43	35
Natural seeps	12	10
Off-shore production	< 2	1
Total	124	100

Source: National Academy of Sciences, Petroleum in the Marine Environment, January 1975. Based on data from Table 1-5, p. 6, as presented in National Petroleum Council, *Ocean Petroleum Resources*, 1975, Table 16, p. 56.

oceans. Given the increase in off-shore petroleum exploration and production in the future, this percentage may increase. Also, on a regional basis, this percentage may be higher than shown because the scope for comparison will be smaller.

The small percentage attributed to exploration and production appears to suggest that it may not be a problem. But damage to the environment is a problem. It becomes a more serious problem because marine pollution, by nature, is not conducive to being contained in or confined to an area by simply defining territories for control responsibility or jurisdiction.

To begin with, very little is known about the impact of pollution because very little is known about the biological resources which are unused or unexploited in the deeper portions of the continental shelf and slope.[60] The extent of these resources must be understood before the impact of oil and gas development on the oceans can be properly assessed.

One of the main impacts in South-East Asia, of course, is on the marine life that is both the source of protein nutrients for most of the population and a source of livelihood for a significant number of small-time fishermen. There could also be damage to the habitats of wildlife. For example, although the actual damage done to the wetlands in off-shore producing states like Louisiana and Texas is not yet clear, there is evidence that petroleum development in the coastal zone and channels dug for pipelines from off-shore fields have caused considerable damage to the salt marshes off Louisiana.[61]

Some other impacts are indirect. Citing again the example of Louisiana, accelerated population growth that results from the attraction of industries to the oil producing zone implies the urbanization of the coastal zone. Off-shore activity also requires on-shore support to service platforms and construct rigs and platforms. Other industries are also attracted by this growth in population and economic activity. All this affects the habitats of birds and animals, marine and wildlife species.

As society increasingly builds environmental quality standards into the social welfare calculus, the environmental impacts of an economic activity such as petroleum resource development cannot be ignored. Any environmental deterioration related to such development as well as international conflicts that may result from transfrontier pollution would offset some of the economic gains from indigenous petroleum production. Even without including the possibility of conflicts, environmental impacts may still result in negative welfare effects if income distributions are altered quite drastically as *a result. The trade-offs that a society faces and must deal with are discussed in somewhat more depth in the following section.

Economic Welfare Implications of Conflict Situations

Economic analysis from the time of Adam Smith has been concerned with the question of optimal resource allocation. The economic problem confronting every nation is that its available resources are scarce relative to the demands of its population and its desire to improve well-being. If these scarce resources are used efficiently, that is, allocated to higher-valued uses, then the well-being of the population will be relatively higher than if either private individuals or government employed such resources in lower valued uses, or used them wastefully. The welfare[62] of a community is the result of aggregating (in some fashion) the welfare of that community's individual members, or their utilities, or satisfaction. The level of individual satisfaction is presumed to depend on his own consumption of goods and services, and enjoyment of the environments he is exposed to, and that level of consumption is, to a certain degree, related to his income.

Discussion of a policy objective to achieve optimal resource allocation presupposes certain social goals. Alternative allocations can be ranked or evaluated by two basic welfare criteria for evaluating policy: *efficiency* and *equity*. It is easier for an economist to define the first than the second. A situation is said to be efficient if it is not possible to rearrange the components so as to benefit one person without harming any others. This economic equation for efficiency is also referred to as the *Pareto optimum*. A Pareto optimum is not a unique allocation of resources. There are many Pareto optima, depending on the initial stock of goods and the initial distribution, so that there are many different Pareto optimal resource allocations consistent with different distributions of satisfactions among the individuals of a society. Economics itself is unable to specify the objective welfare function of a society that would attain that society's welfare maximum, or the maximum of aggregated satisfactions. The Pareto equation cannot be used as a guide to social policy but is only important as a 'preliminary to the determination of a genuine social maximum in the full sense'.[63] To say that a system is Pareto efficient is not to say that it either is or is not equitable. All it says is that no resources would be wasted and that no further improvements could be made without making someone worse off. Economics can suggest ways of achieving efficient states. It can try to describe the equity considerations involved in any suggested policy. But the specific equity goal cannot be decided on economic grounds, for they involve personal and philosophical values. The total welfare objective is, therefore, a political decision and must ultimately be decided by a political process.[64]

It is clear from the above sketch that a nation's desire to seek optimum welfare in terms of optimum satisfaction from the consumption of goods may not be separated from optimum utilization of its resources. Given the value of petroleum to the economic growth of each nation, optimum access to such resource would be part of the welfare-maximizing objective of a nation owning such resources. The ownership system in South-East Asia has allocated all primal rights to off-shore petroleum resources to coastal states rather than to individuals (individuals can contract for such rights from the state). Thus, the timing of the development of off-shore petroleum resources, in so far as decision-making is concerned, is dependent on the state. But the ability of the state to plan such development is dependent on its access to such resources. In the face of disputes over use of these resources, such planning cannot effectively be made. Since petroleum resources are capital resources that are inputs in the productive process, their utilization is an economic variable that is determined in a specific case by an economic decision. But in the general case their utilization is dependent on national policy.

How then can government policy relative to the utilization of a nation's off-shore petroleum resources be evaluated? With the national goal of welfare maximization in mind, benefit-cost analysis can be an analytical tool for a systematic approach to the evaluation of the economic variables that enter into the welfare maximization calculus for this resource. The trade-offs may then be assessed.

Benefit/Cost Analysis in Government Decision-making Relative to Off-shore Petroleum Resources

The methodology for applying benefit/cost analysis to government decision-making relative to the development and utilization of national petroleum resources is similar to that for private decision-making. First, the social value of the project must be estimated. Second, the social costs of operations over time must be identified. Third, the stream of social benefits that flows from the project must also be estimated. Fourth, with these data determined, the social rate of return must be estimated to determine the present value of the net benefits for each of the alternatives under consideration. When the choices are identified, benefits and costs associated with each choice must be estimated. The government would take the decision that would maximize the present value of net social benefits.[65]

A government involved in a dispute over ownership rights of petroleum resources on the continental shelf of South-East Asia may have several alternatives. Basically, however, they may be reduced to two. The first alternative is to aim to take control over the maximum possible area size through protracted negotiations (we discard the use of force altogether). The second alternative is to concede some parts of the territory in question and accept a smaller area size in exchange for a quicker negotiation period. Disputes arising from accumulations dissected by national boundaries or arising from environmental problems give rise to similar alternatives.

The value of petroleum to South-East Asian countries and the potential supply of indigenous resources have already been discussed in the first two chapters. The micro- and macro-economic variables that form some of the bases for evaluating the costs and benefits of conflicts versus cooperation have also been identified in the preceding sections of this chapter. Which alternative will best achieve the welfare-maximizing goal of a nation will be dependent on the present value of petroleum resources under each alternative, and that present value will be dependent on the accepted social rate of return. Since the welfare-maximizing goals of a resource-owning nation are analogous to the profit-maximizing objectives of a firm, the calculation of this social rate of return may be explained, first, in terms of a simplified discussion of capital theory at the micro level,[66] translated later to a societal level.

The value of a petroleum deposit to a producer is the present value of future sales from it, net of extraction costs. Therefore, its value or net return must be growing at a rate equal to the rate of interest, that is, the marginal profit (or revenue minus cost on the incremental unit of output) must be growing like the rate of interest. The resource can produce revenue for the developer today or at some time or series of times in the future. To get all his revenue today, he could sell the deposit. If he decides to produce it over a period of time, he can do so over a short or over a long period of time. And so the developer faces the problem of the *rate* at which he will produce the stock to maximize his returns over the life of the deposit. He could drill one well into the reservoir or several. If he drilled several, he could produce the deposit quickly and push gross returns closer to the present. But to do so would cost him more than if he were to drain the reservoir with one well over a longer period of time. The developer's economic problem is to find that combination of rate of output and time frame which would yield him the maximum net benefits in present value terms over the life of the investment on producing the deposit. This involves a consideration of output, costs, prices, and the interest rate, as well as an estimation of the marginal net return today and for each future period.

Where the price is not known, a price is assumed and multiplied by a declining production profile to yield an income stream. Each phase of the income stream is discounted by a rate that would equate the sum of the present value of the anticipated output to the investment expenditure required to install the producing equipment. The rate of discount so estimated is then compared to the firm's cost of capital, which is defined as either the cost of borrowing money, or the opportunity cost or the return from the best alternative use of that money.[67]

If the net return were to rise too slowly, production would be pushed nearer to the present in time and the deposit would be exhausted quickly, because the developer would not wish to hold the petroleum in the ground and earn less than the going rate of return. The petroleum deposits are a poor way to hold wealth under the circumstances. If net returns are expected to rise very fast, petroleum deposits would be an excellent way to hold wealth, and the developer would delay production.

Thus, theoretically, if the current rate of interest is less than the discounted expected return for future periods, the rate of production should be slowed down to obtain a larger future return. In effect, this means slowing down investment costs, and adjusting the rate of production until the discounted marginal net returns in each future period are equal to the current interest rate and to each other.

This discussion for one producer can be applied conceptually to a country in relation to its petroleum deposits in the off-shore area.[68] The difference is that social benefits and social costs cover a broader spectrum, and quantification must be extremely imprecise. For petroleum resources, the social discount rate is the time rate of change in the present value of the social marginal utility of consumption of this resource. This value depends on the social rate of time preference (that is, the trade-off rate between the costs of foregone present consumption and anticipated larger future benefits), the growth rate of per capita consumption of petroleum, and the elasticity of the marginal utility of per capita consumption. The social discount rate is not, in general, revealed either in free market interest rates or in the marginal productivity of capital, unless the rate of investment is optimal.[69] In estimating the value of resources to a community, shadow prices must be imputed when market values are either not available or are assumed to be irrelevant because of the divergence between private and social returns.

The social rate of discount is usually lower than the market rate, because as a whole a country has a weaker preference for current consumption or return on investments than an individual. Society will tend to look ahead in terms of future generations, whereas the individual will tend to look in terms of its own life span.

The social rate of discount is a tricky question, and its choice involves political judgements. It is dependent on the polity and the priorities it sets for any specific project. For a country with various energy resources capable of immediate development and utilization and which exceed current or short-term needs, the social rate of discount for petroleum may be low. A country unwilling to risk heavy dependence on a foreign source today may have a high social rate of discount. Nevertheless, it is possible to apply a test of economic efficiency and to ask whether or not the costs of choosing a situation are minimal relative to the benefits.

Efficiency in the development and use of natural resources is a relative term, given the conditions existing in a place as opposed to global efficiency. What may not be efficient in the global context may be efficient in the light of the particular economic conditions as well as non-economic constraints and social priorities of a nation. The more open the economy, the more vulnerable it is to external influences. When its foreign dependence is on a resource that is a vital input in production processes, the vulnerability translates into a question of whether the economy grows at a faster or slower rate than desired, all other things held unchanged. Security in the enjoyment of one's standard of living, reduction of unwanted risk in the attainment of national economic goals, perceived justice in the distribution of income, etc. — these are goals that may dictate a pattern of development and consumption of certain resources in one country in a manner different from those desirable in another country. As

Cooper put it, 'If we are to elect an inefficient solution to allocation of resources, we should be sure that we get a great deal for it in terms of our other social objectives'.[70] This is tantamount to measuring social benefits versus social costs, rather than a strict accounting of direct costs and benefits.

Social Opportunity Costs of Disputes

In evaluating a public investment project, the social opportunity cost of that project is the value to society of the next best alternative use to which the resources employed in the project could have been put. The appropriate measure of that social opportunity cost is the discounted value of the consumption stream that would have occurred had the project not been undertaken.[71] The opportunity cost to a nation of disputes related to petroleum resources may be measured in terms of the next best alternative use of the resources employed in settling the dispute, the cost of consuming an alternative fuel in the short or medium terms, because of non-access to all or part of its petroleum resources, and the costs of arriving at an agreement or of cooperation.

The first type of costs comprise the costs of litigation — the employment of human and other resources to settling the dispute rather than to activities that would add to the national economy's output of real goods and services. The costs would be weighed against the long-term future gains that are expected to be derived from a favourable settlement of the dispute. Those gains are related to the second type of costs.

Some of the economic costs associated with the second type may be measured in terms of the economic variables already identified earlier in this chapter. To recapitulate at this point, petroleum development results in benefits and costs related to foreign exchange (both in import costs or export earnings), the availability of petroleum to the economy, fiscal earnings from petroleum production, employment and income — direct and indirect — arising from off-shore petroleum production, and income distribution. There are also environmental effects that need to be included in the welfare calculus. The benefits or costs related to these variables in the process of developing one's petroleum resources are dependent on the friendly or unfriendly situations that arise between governments, and how such situations impact on the variables.

The third type of costs include the effects of cooperation versus non-cooperation on any of the following areas: climate for foreign investments; research costs; common pool operations; pollution control. The long-term impacts of gains or losses in these areas require some elaboration of these topics.

There are many factors that affect *investment* in exploration and development of petroleum resources. They will not be discussed here. It might be pointed out, however, that making the best use of the region's petroleum resources calls for optimum economic development and extraction, given the appropriate cost/price relationships and the national economic goals that dictate such production; it therefore requires optimum conditions favouring investment, among other things.

If the climate for investment (that is, national policies and regulations affecting investment and exploration) is favourable, and if no uncertainty on territorial limits exists, the only factors that are left to the decision-maker are technical and cost/price factors. Any national disagreement on limits would deter some investments because of the uncertainty it creates about property rights. Thus, some major international oil companies have stayed away from the Spratly Islands, even though efforts have been made by the governments of the Philippines and Vietnam to attract exploration in the area. It will be recalled that none of the eleven blocks in the Gulf of Thailand subject to counter-claims drew bids when offered for exploration/exploitation.[72] The effect of hesitance or delay in active exploration in the area translates into delay in producing oil from the region. Although there is no guarantee of discovery in any specific area it means that much longer to wait to discover and produce where production is possible. For the country concerned, that means, among other things, a greater drain on foreign exchange in the short and medium terms.

The benefits of *cooperation in research* can be explained by a simple example of a two-person relationship, to show that individuals with purely selfish motives can mutually benefit from cooperation and exchange. This two-person example can be extended to the national level. Granted, A owns a tract of land which he suspects has petroleum. He hires a geologist/geophysicist to study the rock formations in the area. Since it is generally necessary to understand the geology of contiguous areas to come to any good conclusions on the specific tract of interest, he also has the contiguous areas studied.

Part of this contiguous area is adjacent to another tract of land owned by B and on which B would like to explore for petroleum. B also has a geologist/geophysicist survey his land and the contiguous area. Both would like to know what each other's survey has shown but each does not want to give information about the contiguous area freely for fear one will know more than the other, and so both guard their information jealously.

There are costs — in time, effort, and money — in trying to protect information. There is a cost also in not learning all there is to learn about something. The choices facing A and B are (1) not to trade and only to protect information; (2) not to trade and at the same time to try to steal information; and (3) to trade information. Obviously, the third would be the most mutually beneficial arrangement.

Since this is sharing of information that has no relation to price, it cannot be construed as collusion to increase market power. Rather, there is the societal benefit in the minimization of risk in drilling into the unknown, and in maximization of the benefits from research. The benefits of cooperative research are already recognized in the oil industry both in technology research and in basic data generation — and geological/geophysical surveys fall under the latter.[73]

The difficulty of *confining pollution* resulting from off-shore petroleum development within national boundaries was discussed earlier, because of the free-flowing nature of ocean waters and the mobility of certain marine resources, that is, the 'non-excludability' of such resources. The lower social costs of cooperation and joint pollution control, both in standards and in equipment, need not be detailed here. The cost savings may be viewed, as in the case of cooperation in common pool operations, in terms of savings in reducing allocation of resources strictly to monitoring pollution standards instead of also having to dedicate resources to monitor rival actions by neighbouring countries. Where coordination is necessary, there are also savings in avoidance of duplication of efforts and equipment.

Where national boundaries straddle a pool of hydrocarbons, international cooperation through some form of *unitization agreement* is the best approach to efficient production. If disagreement prevails, there are three sources of societal costs: (1) there is a cost to society when settling disputes arising with regard to possible drainage of oil from one territory to another; (2) there is a loss of resources through inefficient lifting rates, and therefore a loss of revenues to producers and of potential satisfaction of consumers; and (3) there could be higher costs of production and to society if unnecessary investments in wells are made to 'capture' as much of the oil in the sub-surface as possible. In each case, there is a misallocation in the employment of productive resources, and a loss to society in current and future terms. There are also long-run costs to the governments in question from inefficient production rates in the form of lost revenues or output shares, in so far as these are tied to output levels.

1. *Exploration* is the process of locating oil deposits, while *development* consists of finding the horizontal and vertical limits of a petroleum reservoir in any given field located by exploration activities. See M.A. Adelman, *The World Petroleum Market* (Washington, D.C.: Resources for the Future Inc., 1972), pp. 18-19.

2. Risk is distinguished from 'uncertainty' in the degree of the opportunity of loss; the term 'risk' refers to an opportunity for loss, whereas the term 'uncertainty' may be applied to 'factors where the outcome is not certain but where the opportunity for loss is not as apparent as in risk'. This distinction is made in R.E. Megill, *An Introduction to Exploration Economics* (Tulsa, Oklahoma: The Petroleum Publishing Company, 1971), p. 98.

3. Megill, op. cit., p. 99, describes exploration economics as follows: 'Hydrocarbon potential + economics + risk = decision'. The state of technology may be subsumed in the element 'hydrocarbon potential'. It is preferable, however, to identify the influence of technology separately, in view of the distinction between resource potential and actual reserves at any given time period, the latter being highly dependent on existing technology.

4. A similar and expanded discussion of the technical and economic (including the legal and administrative) considerations may be read in the following papers: F.W. van Bilderbeek, 'Offshore exploration drilling and development', in United Nations, *Proceedings of the Third Symposium on the Development of Petroleum*

Resources of Asia and the Far East, Vol. I, 1967, pp. 574-93, and J. Montel, 'Necessary conditions for the development of oil search', in United Nations, *Proceedings of the Second Symposium on the Development of Petroleum Resources of Asia and the Far East*, Vol. II, 1963, pp. 231-4.

5. See P. Leicester, 'Risks and costs of oil exploration and development', United Nations, *Proceedings of the Symposium on the Development of Petroleum Resources of Asia and the Far East*, 1959, pp. 44-8. For discussions of decision methods and yardsticks for measuring uncertainty and risks, see Megill, op. cit., pp. 98ff. and Arthur W. McCray, *Petroleum Evaluations and Economic Decision* (Englewood Cliffs, N.J.: Prentice-Hall, Inc., 1975), Chaps. 3-10.

6. ESCAP, CCOP, *Report of the Eleventh Session* (1974), p. 37, for the Indonesian data. The information on the U.S. success ratio are from data of the American Association of Petroleum Geologists, American Petroleum Institute, and *Oil and Gas Journal*, presented in Corazón M. Siddayao, 'The natural gas shortage: causes, solution', unpublished report submitted to the George Washington University, Department of Economics, Energy Research Project, December 1974, Table 12. The American Petroleum Institute also shows discovery efficiency in off-shore areas of the United States to average 10 per cent or less (see *Quarterly Drilling Report*, 1974).

7. Economist Intelligence Unit, *Quarterly Economic Review— Oil in the Far East and Asia*, June 1975, p. 1.

8. Example from Megill, op. cit., pp. 108-9.

9. Allen I. Hammond, 'Bright spot: better seismological indicators of gas and oil', *Science*, Vol. 185, 9 August 1974, p. 515.

10. A *stratigraphic trap* results from a chance phenomenon; a permeable layer is bounded by non-permeable layers, but with no systematic relationship of one to the other. The permeable layer containing the fluid is stopped or pinched out as the layer ends in contact with a non-permeable layer. A *structural trap*, on the other hand, results from a deformation of one or more underground strata, forming domes or dips or faults. The ultimate block to further migration of the hydrocarbons is at the highest point of the deformation. Distinction described in M.A. Adelman, footnote 43, p. 30, and drawn from Frick and Taylor (eds.), *Petroleum Production Handbook* (1962).

11. The 'bright spots' can indicate the presence of materials other than oil and gas. A geologist reports that in wells drilled in Burma and Japan, the zones which, on the basis of 'bright spots', were expected to yield oil or gas yielded coal instead. By experience also, the 'bright spot' technique is only useful for structures that are shallower than 7 000 feet (about 2 100 metres). Beyond that, the fidelity of recording deteriorates. (This information was received in conversations with Allen G. Hatley, General Manager, Cities Service East Asia, in Singapore, 2 September 1976, and with Geologist Salvador De Luna, formerly of Esso Exploration Inc. and now with Bow Valley Exploration (Pte) Ltd. in Singapore, 12 October 1977.)

12. Discussion of the distinction is based largely on R.J. Kalter, W.E. Tyner, and T.H. Stevens, *Atlantic Outer Continental Shelf Energy Resource: An Economic Analysis* (Cornell University, Department of Agricultural Economics: Ithaca, New York, November 1974), mimeographed, A.E. Res. 74-17, pp. 18-20.

13. Megill, op. cit., pp. 22-3.

14. Paul Weaver, 'Variations in history of continental shelves', *Bulletin of the American Association of Petroleum Geologists*, Volume 34 (1950), cited in Kalter, op. cit., p. 19.

15. See U.S., Council on Environmental Quality, *OCS Oil and Gas—An Environmental Assessment* (Washington, D.C.: U.S. Government Printing Office, 1974), Vol. 1, Chapter 4, pp. 56-64.

16. Seismic data are obtained by bouncing sound waves off the sea bottom to obtain a profile of sub-surface formations. Propane-oxygen guns and high-powered oscillators rather than explosives generate the sound waves. Bottom sampling and coring are used to take rock samples of the ocean floor and sub-surface. *Coring* refers to taking a rock formation sample by drilling a shallow hole; it is used for identifying unconsolidated sediments on the ocean bottom's sub-surface. See *OCS Oil and Gas*, pp. 56-7.

17. Barges are frequently used for shallow water drilling, while drillships are often used to drill in waters deeper than 500 feet or 150 metres. Drillships maintain their location with either anchors or dynamic positioning propellers to compensate for movement. A *jackup* or self-elevating platform has buoyant hulls. When the platform reaches the site, the legs are jacked downward to the ocean floor and the platform deck is raised above the sea surface. A *semi-submersible* drilling platform is similar to a floating drilling rig and is supported by either displacement hulls or large caissons. Semi-submersibles are used extensively for deep sea drilling. See *OCS Oil and Gas*, pp. 57-8.

18. See *Annual Report, 1975* of Oriental Petroleum and Minerals Corporation (Manila).

19. Chase Manhattan Bank, *Capital Investments of the World Petroleum Industry* (New York: December 1975), Schedule 4.

20. Thomas R. Wartelle and Griff C. Lee, 'Fixed platform design for South East Asia', paper presented at the Society of Petroleum Engineers session, Offshore Southeast Asia Conference, 20 February 1976 (Singapore).

21. Conversation with Allen G. Hatley, 2 November 1976.

22. See Robert Dorfman, 'The value of offshore oil', in Edward J. Mitchell, editor, *The Question of Offshore Oil* (Washington, D.C.: American Enterprise Institute for Public Policy Research, 1976, pp. 5-16); National Petroleum Council, *Ocean Petroleum Resource* (Washington, D.C., 1975), pp. 20-3; and American Petroleum Institute et al., *Joint Association Survey of the U.S. Oil and Gas Industry* (Washington, D.C., 1975).

23. Dorfman, op. cit., p. 6.

24. Ibid.

25. Water depths of 50-100 feet (15-30 metres). Conversation with Allen Hatley, 2 November 1976.

26. See World Bank, *Energy and Petroleum in Non-OPEC Developing Countries*,

1974-1980, Staff Working Paper No. 229 (Washington, D.C., February 1976), Tables 10 and 11.

27. *Asian Wall Street Journal* (Hong Kong), 3 November 1976, p. 4, 'Big oil vs. new Malaysian nationalism'.

28. *Petroleum News Southeast Asia*, August 1976, p. 16.

29. U.S., Federal Energy Administration, *The Relationship of Oil Companies and Foreign Government* (Washington, D.C.: U.S. Government Printing Office, June 1975), p. 116.

30. Kalter, p. 3, and Raymond F. Mikesell, *Foreign Investment in Petroleum and Mineral Industries*, Resources for the Future, Inc., (Baltimore, Md.: The Johns Hopkins University Press, 1971), Chapter I.

31. *Petroleum News S.E.A.*, August 1976, p. 14, 'Brunei's billions'.

32. U.S., Office of the President, *International Economic Report of the President* (Washington, D.C.: U.S. Government Printing Office, 1976), p. 5.

33. Weir M. Brown, *World Afloat: National Policies Ruling the Waves*, Essays in International Finance, No. 116, May 1976 (Princeton, N.J.: Princeton University, 1976), p. 6.

34. See Table 4, 'Some implications of rising trend in petroleum price for developing countries, 1970-1985', 20 December 1973, Document SecM73-769.

35. These estimates were in 1973 dollars and the 1974 actual costs included some inflation factors. The differentials would still be slight.

36. See Wouter Tims, 'The developing countries', in Edward R. Fried and Charles L. Schultze, editors, *Higher Oil Prices and the World Economy* (Washington, D.C.: The Brookings Institution, 1975), pp. 169-95; Gerald A. Pollack, *Are the Oil-Payments Deficits Manageable?*, Princeton University, Essays in International Finance, No. 111, June 1975; Thomas D. Willett, *The Oil-Transfer Problem and International Economic Stability*, Princeton University, Essays in International Finance No. 113, December 1975; Andrew D. Crockett and Duncan M. Ripley, 'Sharing the oil deficit', in *Finance and Development*, Vol. 12, No. 4 (December 1975), pp. 12-16. See also *Energy and Petroleum in Non-OPEC Developing Countries, 1974-1980*, World Bank Staff Working Paper No. 229, February 1976.

37. Advocates of the monopoly pricing practices of the OPEC cartel have argued that the marginal cost of producing oil in the Middle East for purposes of world crude oil pricing is not actual extraction cost, but the marginal cost of alternative supplies — the cost of the last barrel of crude produced in the North Sea, in Alaska, and other higher-cost fields, as well as the cost of synthetic fuels from coal. See J. Amuzegar, 'The economic rationale for OPEC's price increase in 1975 and beyond', p. 28, in E. Anthony Copp, editor, *World Petroleum, the Economics of Current Pricing and Supply Policies*, Seminar sponsored by Salomon Brothers, London, England, 20-21 November 1975 (New York City: Salomon Brothers, 1976). This argument stretches the concept of marginal cost pricing.

Societal efficiency demands that the resources that can be produced in the least costly manner be consumed first; and as such relatively lower-cost resources are consumed, society can move on to higher-cost substitutes. That is precisely why coal lost its dominant place in the energy picture in most places before the oil crisis—it was cheaper to use oil at a given place and time than to use coal. As oil in its conventional form becomes more expensive to produce and use, under a competitive price system man will look for and, no doubt, will find other substitutes that are less costly. Advances in technology and man's ingenuity will take care of that.

38. In discussing possible counter-measures by oil-importing, producing nations against the market power of OPEC, Hendrik S. Houthakker has suggested that such nations impose a common tariff on imported oil to reduce consumption and to protect domestic production. See *The World Price of Oil* (Washington, D.C.: American Enterprise Institute for Public Policy Research, October 1976), pp. 29-34.

39. Willett, op. cit., p. 4.

40. Martin Woolacott, 'Brunei: a British anachronism', *Asian Wall Street Journal*, 9 November 1976, p. 4.

41. Lim Joo-Jock, 'Brunei: prospects for a "Protectorate" ', in *Southeast Asian Affairs 1976* (Singapore: Institute of Southeast Asian Studies, 1976), p. 150.

42. Louis Kraar, 'Tomorrow casts a shadow on an Asian Eden', in *Fortune*, August 1976, pp. 192-4; also Woolacott, op. cit.

43. Department of Finance, as published in Economist Intelligence Unit, *Quarterly Economic Review—Indonesia*, Annual Supplement 1976, p. 6. About 40 per cent of Indonesia's GDP comes from the agricultural, forestry and fishing sectors and another 20 per cent from the trade sector. *Statistik Indonesia, 1974/75* showed the share of the mining and quarrying sector to be only 9.3 per cent, in current prices, and 7.0 per cent in 1960 prices, Tables XIV.5 and XIV.6.

44. See Sevinc Carlson, *Indonesia's Oil* (Washington, D.C.: Georgetown University, Center for Strategic Studies, 1976), Chapter III. She discusses the role of oil in Indonesia's economic development. For an analysis of a similar application of petroleum-based revenues to attain economic development objectives, see Donald A. Wells, *Saudi Arabian Development Strategy* (Washington, D.C.: American Enterprise Institute for Public Policy Research, September 1976).

45. Sritua Arief points out that World Bank and Pertamina estimates differ from the U.S. Embassy estimates. He also points out that his own research finds cumulative investments at the end of 1974 to be slightly over US$2 billion rather than slightly under that figure as the table shows. We are, however, interested in changes in the annual level over the 8-year period which he did not provide, and thus have used U.S. Embassy figures. Also, since we are interested in levels, at the magnitudes under consideration, such disparity is not of major consequence for our purposes. See S. Arief, *The Indonesian Petroleum Industry: A Study of Resource Management in a Developing Economy* (Jakarta: S. Arief Associates, 1976), pp. 186-91.

46. Ibid., Tables IV-1 and IV-3, pp. 187, 189.

47. Ibid., Table V-28, p. 282.

48. Ibid., p. 283.

49. Using the formula K/L to measure capital intensity, K was defined as the

value of capital equipment and L as the number of full-time workers. Ibid.

50. Arief estimated the capital intensity of foreign investment projects to be US$6 599:1 and of domestic investment projects to be US$3 740:1. Ibid., p. 284.

51. See U.S. Congress, Senate, Committee of Commerce, *Outer Continental Shelf Oil and Gas Development and the Coastal Zone*, Committee Print (Washington, D.C.: U.S. Government Printing Office, November 1974), p. 46.

52. Ibid., p. 42.

53. This study, *Energy and Petroleum in Non-OPEC Developing Countries, 1974-1980*, was cited in Chapter I. An article 'Energy problems of the non-OPEC developing countries, 1974-80', by Adrian Lambertini, is based on this study, and appeared in *Finance & Development*, Vol. 13, No. 3 (September 1976), pp. 24-8.

54. Lambertini, op. cit., p. 28.

55. See, for example, the case of Venezuela discussed by R. Mikesell, op. cit., Chapter 6.

56. See study by I. Adelman and C.T. Morris, *An Anatomy of Patterns of Income Distribution in Developing Nations*, cited in John H. Adler, 'Development and income distribution', *Finance and Development*, Vol. 10, No. 3.

57. Corazón M. Siddayao, 'Singapore's petroleum sector: a case study of the country's investment growth', in *Foreign Investments in Singapore: Some Broader Economic and Socio-Political Ramifications*, Field Report No. 13 (Singapore: Institute of Southeast Asian Studies, 1977).

58. U.S., Senate, *Outer Continental Shelf*, p. 51.

59. Ibid.

60. Ibid., p. 49.

61. Ibid., pp. 41-2.

62. The term 'welfare' is used in the economic sense of 'well-being'. Economic theory is concerned with prices, output, employment, and income. Welfare economics is concerned with feelings—satisfactions. 'The task of welfare economics is to study the causes of welfare—what would make men happier, and what would not, what people's needs are and how far they can be fulfilled; which needs remain unsatisfied and how far this can be reduced.' A. Radomysler, 'Welfare economics and economic policy', *Economica*, N.S., Vol. 13 (1946), p. 92. Reprinted in *Readings in Welfare Economics* (Homewood, Ill.: Richard W. Irwin, Inc., 1969, for the American Economic Association), pp. 81-94.

63. Kenneth J. Arrow, 'A difficulty in the concept of social welfare', *The Journal of Political Economy*, Vol. 58 (1950), pp. 328-46. Reprinted in *Readings in Welfare Economics*, pp. 147-68.

64. Different models that attempt to incorporate both efficiency and redistribution, with one of the criteria constrained, are discussed in O.C. Herfindahl and A.V. Kneese, *Economic Theory of Natural Resources* (Washington, D.C.: Resources for the Future, Inc., 1974), p. 222-7.

65. See Walter Mead, 'Social benefit/cost analysis of offshore drilling', in Edward J. Mitchell, op. cit., pp. 69-76.
A thorough discussion of benefit/cost analysis and the theoretical problems associated with estimation are given in Chapters 5 and 6 of Herfindahl and Kneese, op. cit.

66. The discussion is based largely on Lovejoy and Homan, op. cit., pp. 16-19 and 89-96.

67. Adelman, op. cit., p. 49.

68. A simple mathematical presentation of the above discussion with special application to society is given in Appendix C.

69. See Charles R. Blitzer, 'On the social rate of discount and price of capital in cost-benefit analysis', World Bank Staff Working Paper No. 144, 1 February 1973, p. 26.

70. Richard N. Cooper, 'The economics of the law of the sea debate', in *Perspectives on Ocean Policy*, Conference on Conflict and Order in Ocean Relations, 21-24 October 1974 (Washington, D.C.: Government Printing Office, undated), p. 144.

71. See Martin S. Feldstein, 'Opportunity cost calculations in cost-benefit analysis', *Public Finance*, Vol. XIX, No. 2 (1964), pp. 117-18.

72. See Chapter IV.

73. See, for example, D.H. Clewell, 'Advantages of cooperative research for the oil industry—what could API's role be?', paper delivered at the American Petroleum Institute Division of Production Meeting, Houston, Texas, 6 March 1972 (manuscript).

VII

Conclusions and Policy Implications

WHILE the demand and supply of petroleum is an economic problem, both are influenced by existing institutions and social policies, and the search for indigenous petroleum resources may not be appropriately viewed outside this setting. Further, when the search for such resources takes exploration to national boundaries, international relations may not be ignored. As Frederick Hayek pointed out years ago:

> If the resources of different nations are treated as exclusive properties of these nations as wholes, if international economic relations, instead of being relations with individuals, become increasingly relations between whole nations organized as trading bodies, they inevitably become the source of friction and envy between whole nations.[1]

The preceding chapters have shown that not only have actual international conflicts already arisen in relation to the existence — discovered or potential — of petroleum resources in off-shore areas of South-East Asia, but the potential for conflict continues to exist beyond the exploration stage into the development and production stage. These conflicts are basically disputes over property rights, whether these be rights over the seabed, the water, or the air. Changes in the property rights system over seabed resources resulting from the new regime of sea rights that have evolved in the past thirty years, as well as conflicting historical claims to ownership, give rise to territorial disputes that need to be resolved at the latest before entering the development stage. The probability that a reservoir may be dissected by two or more national boundaries, accompanied by the non-excludability characteristic of a common pool of petroleum that is being produced, is one potential source of disputes in the producing stage. The non-excludability characteristic of water and air resources constitute other potential conflict areas on environ-mental grounds in the production, transportation, and marketing of petroleum.

Our main concern in this study is why the settlement of ownership disputes related to the development of petroleum resources is im-portant, particularly disputes that restrict access to such resources, for this problem may not be considered separately from a nation's energy policy within the framework of its economic development programme. The related economic policy questions essentially in-volve two cost/benefit considerations. First, what is the cost of allo-cating resources to the disputes versus the beneficial results to be gained from whatever the positive outcome, whatever the specific objective might be? Second, and more important, what is the cost to a nation's economic well-being to be denied access to its petroleum resources at a certain point in time? The first question is more empirical than the second. In both cases, however, generalizations are possible. Both questions centre on the problem of trade-offs, where a choice must be made between two desirable objectives whose maxima cannot be obtained simultaneously.

The economist's notion of the exchange contract curve may be applied to an analysis of the strategy that a nation might employ in negotiating a dispute area. In this simple model of exchange, the Edgeworth box, two individuals with different marginal values on the goods they are trading have dissimilar indifference curves.[2] A contract curve joins the points of tangencies of the traders' in-difference curves. Two trading individuals negotiate until they reach the contract curve, that is, when they reach a point where they place the same marginal value on the goods being traded, a Pareto opti-mum. Negotiations that represent movement towards the contract curve may be characterized as 'positive sum' in the sense that both parties gain by trading. Negotiations that represent movement from one tangency point to another may be thought of as 'constant sum' or 'zero sum', that is, what one party gains, the other loses.[3]

In disputes over petroleum-bearing areas, negotiations that con-sider only the possibility of ending up with the maximum possible amount of area (resources) might be thought of as movements along the contract curve, that is, they are zero sum. When the parties take into consideration the opportunity cost of negotiations in terms of time lost, allocation of resources to litigation that could be devoted elsewhere, and the cost to economic welfare of waiting to gain access to the resources, one might consider the process a case of trading and positive sum. That is, both parties might be said to be considering the marginal cost of protracted negotiations versus the marginal benefits to be gained from the resources that might be won (see points A and B in Fig. 9). In this instance, the indifference curve may be treated as a curve of constant net benefits, where units of one set of net social benefits (social benefits minus social costs) may be traded off for units of another set while remaining at the same level of welfare. That is, the net benefits of rights to delayed access are traded off for the net benefits of rights to early access. Either Point A or Point B would be unsatisfactory to one of the parties, because agreement at either point would mean that one of the parties would accept less than what it would prefer to have. For every negotiating point off the contract curve, there exist negotiating points on the contract curve which are mutually advantageous to Country A and Country B. In this particular case, the parties could agree on Point E and thus move to mutually beneficial positions.[4] Such social costs and benefits are economic considerations that should enter the political decision process when negotiating settlement of territorial and other disputes relating to petroleum.

This leads to the second, and more important, of the two economic policy questions—that which deals with the cost to a nation's economic well-being of non-access to potential petroleum resources within its boundaries. Chapter I explained the importance of energy to economic growth, and the importance of petroleum to South-East Asian economies. It showed that all such countries have been dependent on petroleum for almost all of their energy inputs, and that all but three countries are dependent on imported oil, totally or close to it. With monopolistic pricing in the world oil market, oil import bills have become increasingly onerous for countries with deficient foreign exchange reserves; in addition, as Chapter VI showed, they will be costly in terms of resource allocation impacts and would be paid for at the sacrifice of other economic development programmes. Chapter II showed that the potentials for discovering

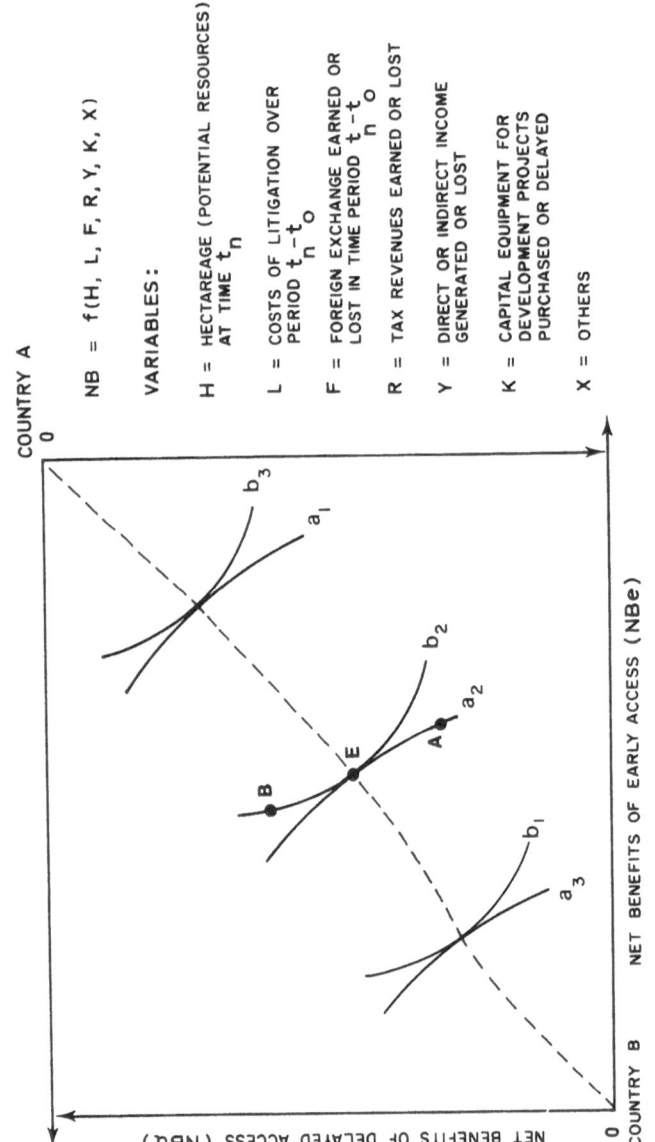

Fig. 9. Trading Off Costs/Benefits of Access to Resources, with Varying Acreages Won and Negotiating Periods

petroleum in South-East Asian seabeds are attractive, both geo-logically and because of current cost/price relationships and existing technology. Given such potential, and the importance of petroleum to South-East Asian economies at present and in the foreseeable future, the pressure to explore for and develop indigenous resources increases with the apparently ever-increasing cost of oil imports.

A recommendation to aim for self-sufficiency in petroleum supply is not to suggest a policy of autarky for its sake, but rather is a rec-ommendation to view energy policy as part of a broader economic development policy. The long-run cost of the inability to obtain re-quired energy inputs for economic development must be weighed against the costs of domestic production. The development of alter-native petroleum supplies constitutes a hedge against arbitrary, uni-lateral price increases from major producing countries that tend to disrupt economic development programmes.

The economic impacts of development of domestic petroleum supply were further explored in Chapter VI. That chapter identified, first, the economic variables that needed to be taken into account in bringing this potential supply into production, and second, those upon which such production would impact. It showed that the in-vestments required to produce the desired supply were huge and that the technology involved is sophisticated. Thus, in general, South-East Asian nations would be relying on imported capital to search for and produce their petroleum resources. In so far as a climate of stability concerning basic property rights (and the right to earn a reward for engaging in the costs of such search) is crucial to attracting such foreign investments, disputes over boundaries or ownership of the seabed serve as negative factors toward such future petroleum production. Chapter VI also showed that petrol-eum production could have huge indirect impacts on income and employment, both on the national and the regional level. At the same time, however, government policy cannot ignore the fact that in the attempt to be self-sufficient in petroleum supply, a reallocation of scarce investment resources away from other development and social programmes would occur. The question is, therefore, what the net benefits or costs are — a problem of choice.

Given the value of petroleum in terms of promoting economic

welfare, the marginal costs of waiting to obtain a larger territory may be higher than the benefits that might be gained from obtaining the desired territory size, when the present value of petroleum resources to the community's economic welfare is considered. Litigations or confrontations may delay the development and production of such resources, directly through inability of operators to proceed with exploration/development or indirectly through the creation of uncertainty that keeps risk investments away from the region. Therefore, over the long term a misallocation of the nation's resources may seriously prevent it from achieving a higher state of economic well-being.

The prospects for cooperation are by no means dim. The progress that has been made by ASEAN is a positive indicator of the potentials for cooperation, and development projects on a regional basis under the sponsorship of the Committee for Coordination of Joint Prospecting for Mineral Resources in Asian Offshore Areas (CCOP) also show promise for cooperation.[5] The ASEAN Council on Petroleum (ASCOPE), a body organized by oil companies, is another indicator. ASCOPE has primarily been concerned to date with product supply to meet consumption requirements. For example, in October 1976 the Council agreed that nations would share energy supplies in the event of another oil shortage similar to that of the mid-1970s.[6] It has, however, also voiced plans to undertake research that will promote efforts to develop the region's petroleum resources.

The importance of cooperation cannot be over-emphasized, for oil resources have become important enough to make countries willing to risk confrontation to attain ownership of such resources. The disputes and confrontations in the Gulf of Thailand, discussed in Chapter IV, have demonstrated this. The national disagreement and polemics between Greece and Turkey over the activities of the Turkish seismic ship in the Aegean Sea is a recent example.[7] In Asian waters, Philippine drilling in the Reed Bank despite warnings by China, Taiwan, and Vietnam to keep off what each claims to be its territory is another example.

Thus, if the resources are there but are inaccessible because of property rights disputes, the questions each decision maker may ask himself are: How much do we want them? Why? When do we want them? How much are we willing to give up to get what we need and when? What is the social rate of discount? It did not require this study to come to a conclusion that it makes economic sense for nations to cooperate in energy matters to maximize welfare opportunities over the long run. This study, we hope, has demonstrated the stakes involved, and shown that regional cooperation in energy matters in the developing countries of South-East Asia is essential for basic economic reasons.

1.   *The Road to Serfdom* (Chicago, Ill.: University of Chicago Press, 1967), p. 220. Quoted by James Grant (Barron's), 'Commodities and the struggle for wealth', in *Asian Wall Street Journal* (Hong Kong), 6 September 1976, p. 6.

2.   Application of the Edgeworth box to the theory of exchange is given in Nancy Smith Barrett, *The Theory of Microeconomic Policy* (Lexington, Mass.: D.C. Heath and Company, 1974), pp. 101-2.

3.   This explanation takes off from Robert D. Tollison and Thomas D. Willett's proposed Theory of Optimal Negotiations, presented in short form in 'Constitutional mechanisms for dealing with international externalities: a public choice perspective', *The Law of the Sea: U.S. Interests and Alternatives,* pp. 97-8. Thorough discussions of game theory are given in Robert Dorfman, Paul A. Samuelson, and Robert M. Solow, *Linear Programming and Economic Analysis* (New York: McGraw-Hill, 1958), Chapter 15, and in William J. Baumol, *Economic Theory and Operations Analysis* (Englewood Cliffs: Prentice-Hall, 1961), Chapter 18.

4.   The objective function of each negotiating country may be viewed as follows:

$$Max\ W = f(B - C) = f(NB)$$

where W = welfare level, B = benefits, C = costs, and NB = net benefits, and where such costs and benefits are a function of several variables. These variables include hectarage (or potential resources) gained or lost, the costs of litigation, potential foreign exchange earned or lost in time period $t_n - t_o$, potential tax revenues earned or lost, potential direct or indirect income generated or lost, and capital equipment for development projects purchased or delayed. The equation for an indifference curve in Figure 9 may be expressed as follows:

$$\frac{\partial W}{\partial x}\ dx\ +\ \frac{\partial W}{\partial y}\ dy = 0,\ \text{where}\ x = NB_e\ \text{and}\ y = NB_d.$$

At point E, where the indifference curves of Country A and Country B are tangent, their slopes $(-dy/dx)$ are equal. That is, where the countries are at mutually beneficial positions, their trade-off rates are equal:

$\dfrac{W_{Ax}}{W_{Ay}} = \dfrac{W_{Bx}}{W_{By}}$, where W_{ij} is the partial derivative of the welfare function of Country

i with respect to an alternative set of net benefits j.

5.   Singapore, *Trade and Industry Supplement*, February 1976, p. 19.

6.   *Asian Wall Street Journal*, 18 October 1976, pp. 1, 5, 'ASEAN agrees to share fuel in emergency'.

7.   'Dispute over Aegean seabed', by David Tonge, *Guardian Weekly*, 1 August 1976, in *Mirror*, 16 August 1976, p. 2.

Appendixes

U.S. Geological Survey Description of Coastal Categories Used in Table 2.2

Landlocked—country has no access to coastal waters.

Shelf-locked—country has access to seabed areas which have super-jacent waters that are no deeper than 200 metres.

Open shelf—country has access to areas deeper than 200 metres.

Archipelago—indicates that country is an island or islands on a shelf area.

Coastal length—length of coastline facing the sea, exclusive of irregularities.

Shelf and margin:
   Continental shelf—defined as the area between the shoreline and the 200-metre isobath.
   Outer edge of margin—generally lies between the 200-metre and the 3 000-metre isobaths.

Off-shore—the term is used in a more restricted sense than by the oil and gas reporting media. Off-shore waters are defined as (1) waters along shores facing high seas, or (2) waters along shores of relatively confined seas through which an international boundary exists.

Bodies of water surrounded by one nation or inland seas are not regarded as off-shore waters under this definition. Example: Lake Maracaibo or the Caspian Sea are not off-shore by this definition.

.

Conversion factors

   1 nautical mile = 1.15 statute miles = 1.85 kilometres

   1 square nautical mile = 1.32 square statute miles = 3.43 square
   kilometres.

Reference: U.S. Geological Survey, *Summary, 1972 Oil and Gas Statistics, Onshore and Offshore Areas, 151 Countries*, Professional Paper 885, 1974, pp. 2-3.

APPENDIX B

Proposed Legal Definitions in the Third Law of the Sea Conference

THE Informal Composite Negotiating Text incorporated amendments suggested and accepted as a result of discussions during the fourth, fifth, and sixth sessions in New York in 1976 and 1977. Below are excerpts pertinent to the topic of development of South-East Asia's petroleum resources but which did not quite fit into the main text.[1] Also included in a few places are short discussions of related issues.

TERRITORIAL SEA AND CONTIGUOUS ZONE

Article 3 provides that the breadth of the territorial sea may be established by a nation to a limit not exceeding 12 nautical miles. In localities where the coastline is deeply indented and cut into, or where there is a fringe of islands in the immediate vicinity of the coast, the method of straight baselines joining appropriate points will be employed (Article 7.1). Where this method results in enclosing as internal waters (defined as 'waters on the landward side of the baseline of the territorial sea') areas which had heretofore been otherwise, the Convention provides for the right of innocent passage through such waters (Article 8.2).

Where the coasts of two states are opposite or adjacent to each other, the limit of the territorial sea of each coastal state will be the median line, every point of which is equidistant from the nearest points on the baselines (Article 15). However, the same article provides that this would not apply to territorial seas already delimited by reason of historic title 'or other special circumstances'.

Among other things, the rights of the state within the territorial sea include the following: (1) conservation of the living resources of the sea, (2) preservation of the environment of the coastal state and prevention of its pollution, and (3) regulation of marine scientific research and hydrographic surveys (Article 21). The coastal state may also require foreign ships exercising the right of innocent passage through territorial waters (including straits) to use certain sea lanes and traffic separation schemes that the coastal state may designate; such restrictions may particularly be directed to tankers, nuclear-powered ships and ships carrying nuclear or other dangerous materials (Articles 22 and 41).

Article 33 extends the limits of the contiguous zone to 24 nautical miles from the baseline from which the breadth of the territorial sea would be measured.

THE HIGH SEAS

The term 'high seas' is defined as all parts of the sea that are not included in the exclusive economic zone, in the territorial sea, or in the internal waters of a state. The high seas would be open to all states, coastal or land-locked (Articles 86 and 87).

The 1958 Convention defines the 'high seas' as the marine area adjacent to the coast of a state where it exercises sovereignty over the sea, the air space over the territorial sea, as well as to its bed and subsoil. (Convention on the High Seas, 1958, Article 1.)

ARCHIPELAGIC PRINCIPLE

The term 'archipelago' is defined as a 'group of islands, including parts of islands, interconnecting waters and other natural features which are so closely interrelated that such islands, waters and other natural features form an intrinsic geographical, economic and political entity, or which historically have been regarded as such'. An archipelagic state may draw straight baselines joining the outermost points of the outermost islands and drying reefs, from which its territorial waters, contiguous zone, exclusive economic zone, and continental shelf would be measured. These baselines may not exceed certain limits, and waters within these baselines would be considered internal waters (Articles 46 to 54).

CONTINENTAL SHELF

Article 76 defines the continental shelf of a coastal state as comprising the seabed and subsoil of the submarine areas located beyond the state's territorial sea

... throughout the natural prolongation of its land territory to the outer edge of the continental margin, or to a distance of 200 nautical miles from the baselines from which the breadth of the territorial sea is measured where the outer edge of the continental margin does not extend up to that distance.

The draft articles provide for sovereign and exclusive exploration and exploitation rights over the natural resources on the shelf, that is, the mineral and other non-living resources of the seabed and subsoil plus living organisms belonging to sedentary species on or under the seabed and subsoil. Exploration and exploitation by any party requires express consent of the coastal state (Articles 77 to 81).

Beyond the 200-nautical mile zone, any exploitation by the coastal state would require payment or contribution in kind to the International Authority that would be set up to administer such area beyond national jurisdiction (Article 82).

Delimitation of the continental shelf between adjacent or opposite states will also be made on the principle of equity, applying the median or equidistant line where appropriate (Article 83).

The Continental Shelf Convention did not extend to the coastal state sovereign rights over living resources in the superjacent waters of the shelf beyond territorial waters, but the Convention of Fishing provided that all states would have the right to engage in fishing on the high seas.[2]

THE ECONOMIC ZONE

Article 55 states that the exclusive economic zone is the area *beyond* and adjacent to a coastal state's territorial sea. Article 57 limits this zone to a breadth of not more than 200 nautical miles from the baselines from which the breadth of the territorial sea is measured.

In this zone the coastal state has sovereign rights over exploration, exploitation, conservation and management of living and non-living natural resources of the bed, the subsoil, and the superjacent waters. It also has exclusive rights and jurisdiction with regard to (1) production of energy from water, currents, and winds, (2) scientific research, and (3) preservation of the marine environment (Article 56).

The draft articles also provide for optimization in the use of the zone's living resources and for access by other nations, including land-locked states. (See Articles 62, 69, 70 and 71.)

The boundary of the exclusive economic zone between adjacent or opposite states would be determined, where appropriate, on the principle of equity by application of the medium or equidistant line (see Article 74). Rocks which cannot sustain human habitation or economic life of their own would not be accorded an exclusive economic zone or continental shelf (Article 121).

At deliberations of the United Nations Sea-Bed Committee which met to prepare the Agenda for the Third Law of the Sea Conference, several delegations of coastal states sought to reserve, for their exclusive use, the living and non-living resources of the adjacent sea areas. One of the proponents of the 'exclusive economic zone' concept was Kenya, which submitted draft articles to this effect. Article I stated: 'All states have a right to determine the limits of their jurisdiction over the seas adjacent to their coasts beyond a territorial sea of 12 miles in accordance with the criteria which take into account their own geographical, geological, biological, ecological, economic and national security factors'.[3] Article II stated that '... all states have the right to establish an Economic Zone beyond the territorial sea ... in which they shall exercise sovereign rights over natural resources for the purpose of exploration and exploitation'. Within such zone the coastal state 'shall have exclusive jurisdiction for the purpose of control, regulation and exploitation of both living and nonliving resources ...'. Article VII fixed the maximum limit of the zone at 200 nautical miles, measured from the baselines for determining the territorial sea, but it left the actual limit pertaining to any country to 'criteria in each region'.

Later versions also followed the same tack, and specified that the natural resources referred to were both renewable and non-renewable, in the seabed as well as in the superjacent waters.[4]

Economic zone of an island. The definition of an island is crucial to the definition of territorial waters for archipelagic states, as well as in relation to the definition of the continental shelf and economic zones of such states. An unrestricted definition would allow a state to claim enormous areas extending outward from tiny specks of rock which could not support habitation. The 1958 Convention on the Territorial Sea defines an island as 'a naturally-formed area of land, surrounded by water, which is above water at high tide' (Convention, Article 10.1). This definition, which also appears in Article 121 of the Composite Text, is not necessarily applicable in defining the continental shelf or economic zone of islands or island groups for the purposes of exploitation of natural resources.

ENCLOSED OR SEMI-ENCLOSED SEAS

States bordering an enclosed or semi-enclosed sea (defined as a gulf, basin, or sea surrounded by two or more states and connected to the open seas by a narrow outlet or consisting primarily of the territorial seas and exclusive economic zones of two or more coastal states) are to 'cooperate with each other in the exercise of their rights and duties under the present Convention' (Articles 122 and 123). It is not really clear what their rights are. Landlocked states are given specific rights 'to participate in the exploitation of the living resources of the exclusive economic zones of adjoining coastal states on an equitable basis' (Article 69). Developing coastal states

... which are situated in a sub-region or region whose geographical peculiarities make such states particularly dependent for the satisfaction of the nutritional needs of their populations upon the exploitation of the living resources in the exclusive economic zones of their neighbouring states and developing coastal states which can claim no exclusive economic zone of their own

have the right to participate on 'an equitable basis' in the exploitation of living resources in the exclusive economic zones of other states in that area (Article 70). In each case, however, the terms and conditions of such participation will be determined through bilateral, sub-regional, or regional agreements among the states concerned. Given the sovereign and exclusive rights accorded the coastal state over the economic zone, the tone of the provisions suggests that the coastal states hold the power to dole out such participation rights (Articles 62 to 68).

The question of closed or semi-closed seas has also become a source of wrangling and of delay in coming to an agreement on the issue of the economic zone. Certain states have insisted strongly on the need for some special arrangements for the seabed within such seas. The concern has related to fishing rights on the ocean surface as well as to resources beneath the sea. Under the Convention on Fishing on the High Seas, the high seas are open to fishing by all nations. The 1958 Convention on the Continental Shelf granted to a coastal state rights over the continental shelf beyond its territorial waters only to resources on the seabed and subsoil. The economic zone concept would extend to the coastal state rights to fishing as well as seabed and subsoil resources to a breadth of 200 nautical miles compared to a maximum territorial sea breadth of 12 miles under the existing laws of the sea. Thus, a country like Thailand, which has a fishing industry that extends to the high seas as currently defined, stands to lose fishing areas should new definitions reduce its capability to fish in the high seas. Although this has nothing to do with oil exploration and development, any delay in arriving at a Convention on the Economic Zone could delay resolution of areas of possible conflicts.

APPENDIX C

Mathematical Explanation of Welfare Maximization in the Utilization of a Society's Petroleum Resources[5]

THE following discussion is intended to explain in simple mathematical form the discussion on cost/benefit analysis and the social discount rate in Chapter VI. The basis of some of the formulations is explained in the section on the rudiments of petroleum accumulation given in Chapter V.

WELFARE MAXIMIZING OUTPUT

The analytical framework presented here incorporates the following factors: the dependence of total ultimately recoverable reserves upon the rate of production, possible control of production decline rates, and risk. The framework is based on the assumption that there is a known quantity of oil in place, S, whose production is constrained by engineering and economic factors. This may be represented by:

$$(1) \quad \int_0^T q(t)\,dt \leq xS$$

where S represents original oil in place, x is a percentage indicating the maximum oil in place that is physically recoverable with existing technology, T is the production time horizon, and $q(t)$ the rate of production. The unknown function $q(t)$ may be operationalized if it is expressed as a function of the initial capacity installed and the production decline rate such that:

$$(2) \quad q(t) = q_o e^{-at},$$

where o is the subscript indicating the present, t is any point in time, q_o represents initial installed capacity,[6] and a represents the rate of decline in production. It will be recalled in Chapter V that total resource recovery is generally a negative function of the rate of production. It may thus be postulated that reserves, S_o, may be expressed as:

$$(3) \quad S_o = xS - \beta q_o e^{-a} - \gamma q_o,$$

where β and γ are physical parameters related to geological conditions, $q_0 e^{-a}$ is the initial rate of production, and q_0 is the installed capacity.

A country may then be said to want to maximize the cumulative output written as:

$$(4) \quad max \int_0^T q_0 e^{-at} dt = S_0 .$$

Equation 4 states that for a given production time horizon, the maximum cumulative output is equal to the magnitude of reserves which may be recovered at a given rate of production and installed capacity.

THE SOCIAL RATE OF DISCOUNT

There will, however, be social costs in the production of the above reserves, and a country will try to maximize the net benefits from such production. The concept will be explained in terms of profit maximization by a producer. On the national level, of course, the benefits from development and production of a nation's petroleum resources go beyond a calculation of the revenue from the sale of such resources. Social costs also encompass more than the direct costs of production, and the social discount rate may not equal the private interest factor. The impacts are discussed in Chapter VI.

Let θ = physical parameter related to initial reservoir conditions

K_0 = initial operating costs per unit of capacity

then

$$(5) \quad q_0 K_0 e^{\theta t} = \text{total operating costs at any point in time.}$$

Unit costs would then be:

$$(6) \quad q_0 K_0 e^{\theta t} / q_0 e^{-at} = K_0 e^{(\theta + a)t}$$

Since total operating costs increase by the value θ through time, but remain constant in any time period regardless of the decline rate, unit costs increase at an exponential rate as production declines through time (Kalter, p. 44).

Notation may be simplified by calling the stream of revenues R and the stream of costs C, as follows:

(7) $\displaystyle\int_0^T pq(t)e^{-at}dt = R$,

where p is price per unit and is given,

(8) $\displaystyle\int_0^T q(t)\,K(t)e^{(\theta + a)t}dt = C$,

where $q(t)$ is installed capacity at any point in time.

Net marginal revenue, or profit, may then be represented by

(9) $R' - C' = V$

where $R' = \partial R/\partial q$, the long run marginal revenue, and $C' = \partial C/\partial q$, the long run marginal costs.

A producer seeks to maximize over time his net marginal revenue. Assuming the engineering and geological conditions mentioned in the first section of this Appendix and knowledge of the size of the reservoir, he achieves his objective if the present value of his discounted net marginal revenue over the long-term equals his net revenue from selling all his rights today. The present value of V is represented as follows:

(10) $V_0 = V_t \cdot \dfrac{1}{(1 + i)}t \cdot (1 + s_t)$

or $\quad V_0 = V_t e^{-it}(1 + s_t)$

where i = the rate of interest for the producer, adjusted for risk and uncertainty

s_t = the fraction of a barrel lost from ultimate recovery and which might have been recovered in time t, for every barrel of production transferred from time t to time o; that is, for every barrel produced beyond the optimum rate.

In the term $\dfrac{1}{(1 + i)}t$, i would be the appropriate positive social discount rate, and may be expressed in the following equation form:[7]

(11) $i = \rho - \dfrac{\dot{U}_c}{U_c}$

where $\jmath$ reflects social time preference, c is per capita consumption, U_c represents the partial derivative of the utility function with respect to c, and $\dot{U}_c$ is a time derivative. In the literature, $\rho \geq 0$, reflecting a preference for the present generation over future generations. The second term on the right half (which will, in general, be negative) may be re-written as follows:

$$(12) \quad \frac{\dot{U}_c}{U_c} = \left\{ \frac{\dot{U}_c}{U_c} \cdot \frac{c}{\dot{c}} \right\} \left\{ \frac{\dot{c}}{c} \right\}$$

where $\dot{c}$ is the time derivative of per capita consumption. The first bracketed expression on the right half of the equation is the elasticity of marginal utility with respect to per capita consumption. This elasticity will always be negative, if the utility function reflects diminishing marginal utility. The second term on the right half is the growth rate of per capita consumption. (Since analysis of the petroleum industry is a partial analysis relative to national growth, this discount rate would reflect only partial opportunity costs and not a general social discount rate.)[8]

The last term in the equation, $(1 + s_t)$, is unique to oil production, and reflects the fact that high rates of production today can actually reduce the total amount of oil that can ultimately be recovered.[9] A society may be quite willing to make this sacrifice if a barrel of oil produced in the future is worth less than a barrel of oil produced today. Higher expected future benefits from production and consumption of the oil would tend to push the production rate up in the future and down today. Thus, the size of i will determine, all other things being equal, what size s a nation is willing to lose.

1.   United Nations, Third Conference on the Law of the Sea, Informal composite negotiating text', Document A/Conf.62/WP.10/Corr. 2, 10 July 1977.

2.   Convention on Fishing and Conservation of the Living Resources on the High Seas, reproduced in U.S. Congress, *Legislation on Foreign Relations*, pp. 1256-62. In the Far East, only Malaysia and Thailand ratified this Convention.

3.   United Nations, General Assembly, *Report of the Committee on the Peaceful Uses of the Sea-Bed and the Ocean Floor Beyond the Limits of National Jurisdiction*, 27th session, Supplement, No. 21 (New York, 1972), p. 180.

4. Document A/Conf. 62/L.4, Working Paper, in United Nations, Third Conference on the Law of the Sea, *Official Records*, Vol. III, p. 81-3.

5. Based on discussion in Chapter V, 'An OCS leasing model', of R.J. Kalter *et al.*, *Atlantic Outer Continental Shelf Energy Resources: An Economic Analysis* (Ithaca, New York: Cornell University, November 1974); on the well-spacing and production-control model in Chapter 4 of W.F. Lovejoy and Paul T. Homan, *Economic Aspects of Oil Conservation Regulation* (Washington, D.C.: Resources for the Future, Inc., by the Johns Hopkins University, 1967); and Chapter 5, 'Investment criteria and time', in O.C. Herfindahl and A.V. Kneese, *Economic Theory of Natural Resources* (Washington, D.C.: Resources for the Future, Inc., by Charles E. Merrill Publishing Company, 1974).

6. The term 'installed capacity' refers to the rate of production that is sustainable for a brief period without further drilling. See Lovejoy and Homan, op. cit., p. 98.

7. The following discussion is taken from C.R. Blitzer, 'On the social rate of discount and price of capital in cost-benefit analysis', World Bank Staff Working Paper No. 144, 1 February 1973.

8. See Oscar R. Burt and Ronald G. Cummings, 'Production and investment in natural resource industries', *American Economic Review*, Vol. 60, No. 4 (September 1970), pp. 576-90.

9. Lovejoy and Homan, op. cit., p. 92.

Bibliography

Economic Theory, Analysis and Policy

ADELMAN, M.A., *The World Petroleum Market*, Washington, D.C.: Resources for the Future, 1972.

ADLER, JOHN H., 'Development and income distribution', *Finance and Development*, Vol. 10, No. 3 (September 1973), pp. 2-5.

ARIEF, S., *The Indonesian Petroleum Industries: A Study of Resource Management in a Developing Economy*, Jakarta: Sritua Arief Associates, 1976.

ARROW, K.J. and SCITOVSKY, T. (eds.), *Readings in Welfare Economics*, Homewood, Ill.: Richard W. Irwin, Inc., 1969.

BAUMOL, W.J., *Economic Theory and Operations Analysis*, Englewood Cliffs, N.J.: Prentice-Hall, 1965.

BLITZER, C.R., 'On the social rate of discount and price of capital in cost-benefit analysis', World Bank Staff Working Paper No. 144, 1 February 1973.

BROWN, W.M., *World Afloat: National Policies Ruling the Waves*, Essays in International Finance, No. 116, Princeton, N.J.: Princeton University, May 1976.

BURT, O.R. and CUMMINGS, R.G., 'Production and investment in natural resource industries', *American Economic Review*, Vol. 60, No. 4 (September 1970), pp. 576-90.

COPP, E.A., *World Petroleum: The Economics of Current Pricing and Supply Policies*, a seminar sponsored by Salomon Brothers, London, 20-21 November 1975, New York: 1976.

CROCKETT, A.D. and RIPLEY, D.M., 'Sharing the oil deficit', *Finance and Development*, Vol. 12, No. 4 (December 1975), pp. 12-16.

DASGUPTA, A.K. and PEARCE, D.W., *Cost-Benefit Analysis: Theory and Practice*, Macmillan Press, 1972.

DORFMAN, R., SAMUELSON, P.A., and SOLOW, R.M., *Linear Programming and Economic Analysis*, New York: McGraw-Hill Book Company, 1958.

DORFMAN, R. and DORFMAN, N.S., *Economics of the Environment, Selected Readings*, New York: W.W. Norton and Company, 1972.

EPPLE, D., *Petroleum Discoveries and Government Policy*, Cambridge, Mass.: Ballinger Publishing Co., 1975.

FELDSTEIN, MARTIN S., 'Opportunity cost calculations in cost-benefit analysis', *Public Finance*, Vol. XIX, No. 2 (1964), pp. 117-18.

FRIED, E.R. and SCHULTZE, C.L. (eds.), *Higher Oil Prices and the World Economy: The Adjustment Problem*, Washington, D.C.: The Brookings Institution, 1975.

GONZALO, L., 'Philippine energy: supply and demand situation', draft report, PREPF Research Note No. 56, Manila: Development Academy of the Philippines, May 1976.

HABERLER, G., *Oil, Inflation, Recession and the International Monetary System*, Reprint No. 45, Washington, D.C.: American Enterprise Institute, 1976.

HERFINDAHL, O.C. and KNEESE, A.V., *Economic Theory of Natural Resources*, Washington, D.C.: Resources for the Future, Inc., 1974.

HOUTHAKKER, H.S., *The World Price of Oil*, Washington, D.C.: American Enterprise Institute, 1976.

KALTER, R.J., TYNER, W.E., and STEVENS, T.H., *Atlantic Outer Continental Shelf Energy Resources: An Economic Analysis*, Ithaca, N.Y.: Cornell University, Department of Agriculture Economics, November 1974 (A.E. Res. 74-17).

KULLER, R.C. and CUMMINGS, R.G., 'An economic model of production and investment for petroleum reservoirs', *American Economic Review*, Vol. 64 (1974), pp. 66-79.

LAMBERTINI, A., 'Energy problems of the non-OPEC developing countries, 1974-80', *Finance and Development*, Vol. 13, No. 3 (September 1976), pp. 24-8.

LOVEJOY, W.F. and HOMAN, P.T., *Economic Aspects of Oil Conservation Regulation*, Baltimore, Md.: Resources for the Future, 1967.

McCRAY, A.W., *Petroleum Evaluations and Economic Decisions*,

Englewood Cliffs, New Jersey: Prentice-Hall, Inc., 1975.

MEGILL, R.E., *Exploration Economics*, Tulsa, Oklahoma: The Petroleum Publishing Company, 1971.

MIKESELL, R.F., *Foreign Investments in the Petroleum and Mineral Industries*, Washington, D.C.: Resources for the Future, Inc., 1971.

MITCHELL, E.J. (ed.), *The Question of Offshore Oil*, Washington, D.C.: American Enterprise Institute, 1976.

Organization for Economic Co-operation and Development, *Oil: The Present Situation and Future Prospects*, Paris, 1973.

PARRA, A.A., 'Some considerations on the demand and supply of petroleum in the seventies in developing countries', paper read at Interregional Seminar on Petroleum Refining in Developing Countries, New Delhi, 22 January—3 February 1973, United Nations Document ESA/RT/AC.5/1.

POLLACK, G.A., *Are the Oil-Payments Deficits Manageable?*, Essays in International Finance, No. 111, Princeton, N.J.: Princeton University, June 1975.

ROBICHECK, E.W., 'The payments impact of the oil crisis: the case of Latin America', *Finance and Development*, December 1974, Vol. 11, No. 4, pp. 12-17.

SAMUELSON, P.A., *Foundations of Economic Analysis*, Cambridge, Mass.: Harvard University Press, 1947.

SCHURR, S.H. (ed.), *Energy, Economic Growth and the Environment*, Washington, D.C.: Resources for the Future, 1972.

SCHYDLOWSKY, D.M., 'Benefit-cost analysis of foreign investment proposals—the viewpoint of the host country', Economic Development Report No. 170, Cambridge, Mass.: Harvard University, Center for International Affairs, June 1970.

SIDDAYAO, C.M., 'The natural gas shortage: causes, solution', unpublished report for the Energy Research Project, George Washington University, Washington, D.C., December 1974.

————, 'Patterns in the utilization of the major energy resources of the United States', unpublished report for the Energy Policy Project, Ford Foundation, Washington, D.C., June 1974.

————, 'Singapore's petroleum sector: a case study of the country's investment growth', in *Foreign Investments in Singapore: Some Broader Economic and Socio-Political Ramifications*,

Field Report No. 13, Singapore: Institute of Southeast Asian Studies, 1977.

SOLOW, R.M., 'The economics of resources or the resources of economics', *American Economic Review*, Vol. 64 (1974), pp. 1-21.

United Nations, Economic and Social Council for Asia and the Pacific, Committee on Trade, 'Implications of the recent changes in the international energy market for developing countries of the ESCAP region', Document E/CN.11/TRADE/L. 254, 20 November 1974.

_____, Department of Economics and Social Affairs, 'World energy requirements and resources in the year 2000', in *Peaceful Uses of Atomic Energy*, New York, 1972.

United States, Congress, Joint Economic Committee, *Resource Scarcity, Economic Growth, and the Environment, Hearings*, Washington, D.C.: U.S. Government Printing Office, 1974.

_____, Senate, Committee on Interior and Insular Affairs, *Outer Continental Shelf Policy Issues, Hearings*, Washington, D.C.: U.S. Government Printing Office, 1972.

VAN MEURS, A.P.H., *Petroleum Economics and Offshore Mining Legislation*, Amsterdam: Elsevier Publishing Co., 1971.

WALTER, I., *International Economics of Pollution*, London: Macmillan Press Ltd., 1975.

WELLS, D.A., *Saudi Arabian Development Strategy*, Washington, D.C.: American Enterprise Institute, September 1976.

WILLETT, THOMAS D., *The Oil-Transfer Problem and International Economic Stability*, Essays in International Finance No. 113, Princeton, N.J.: Princeton University, December 1975.

World Bank, *Energy and Petroleum in Non-OPEC Developing Countries, 1974-1980*, Staff Working Paper No. 229, Washington, D.C.: February 1976.

_____, 'Some implications of rising trend in petroleum price for developing countries', Document SecM73-769, Washington, D.C.: 20 December 1973.

Law of the Sea and Territorial Conflicts

AMACHER, R.C. and SWEENEY, R.J. (eds.), *The Law of the Sea: U.S.*

Interests and Alternatives, Conference sponsored by American Enterprise Institute, Washington, D.C.: 1976.

Asian Wall Street Journal (Hong Kong), 'Dispute over islands in South China Sea is fueled by oil find', 9 September 1976, pp. 1, 9.

_____, 'Greece fails to gain on oil surveying by Turkey in Aegean', 13 September 1976, p. 6.

_____, 'Sea law session ends as members divide on progress', 20 September 1976, pp. 1, 7.

_____, 'Territorial waters of Burma extended', 12 April 1977, p. 6.

Brunei Government, Proclamation over Continental Shelf, 30 June 1954.

Bulletin Today (Manila), 'No easy solution on Spratly, Paracel problem', 6 May 1976, p. 3.

Burma, People's Congress of, Territorial Sea and Maritime Zones Law (Pyithu Hluttaw Law No. 3 of 1977), 28 March 1977.

Cambodia, Royal Government of, 'Declaration du Gouvernement Royal en date du 27 Septembre 1969 relative à la mer territoriale et au plateau continental du Cambodge'.

COOPER, R.N., 'The economics of the law of the sea debate', *Perspectives on Ocean Policy*, Conference on Conflict and Order in Ocean Relations, 21-24 October 1974, Washington, D.C.: U.S. Government Printing Office, 1974, pp. 143-65.

Economic Bulletin (Singapore), 'Spratly islands', July 1975, p. 26.

Far Eastern Economic Review (Hong Kong), 'A new bid for order on the seas', 2 July 1976, pp. 71-2.

_____, 'Manila probes a sensitive spot', 28 May 1976, p. 115.

_____, *Asia 1976 Yearbook*.

GAMBLE, J.K. JR. and PONTECORVO, G. (eds.), *Law of the Sea: The Emerging Regime of the Oceans*, Cambridge, Mass.: Ballinger Publishing Company, 1973.

The Guardian Weekly (Paris), 'Disputes over Aegean seabed', reprinted in *The Mirror* (Singapore), 16 August 1976, p. 2.

HEINZIG, D., 'Disputed islands in the South China Sea: Paracels-Spratlys-Pratas-Macclesfield Bank', Wiesbaden: O. Harrasowitz, 1976.

International Court of Justice, North Sea Continental Shelf Cases, Judgement of 20 February 1969.

LAY, S.H., CHURCHILL, R., and NORDQUIST, M. (compilers), *New Directions in the Law of the Sea*, Vols. I-IV, Dobbs Ferry, N.Y.: Oceana Publications, 1973.

LEE, L., 'The PRC and the South China Sea', *Current Scene*, Vol. XV, No. 2 (February 1977), pp. 1-12.

LOGUE, D., SWEENEY, R., and WILLETT, T., 'Optimal leasing policy and the development of outer continental shelf hydrocarbon resources', *Land Economics*, Vol. LI, No. 3 (August 1975), pp. 191-207.

LUARD, E., *The Control of the Sea-Bed*, London: William Heinemann Ltd., 1974.

MULLOY, P.A., 'Political storm signals over the sea', *Natural History*, 1973.

New Nation (Singapore), 'Vietnam stamps its claim', 26 July 1976, p. 7.

New Philippines (Manila), 'Government states position on imbroglio over isles', February 1974, pp. 6-9.

ODA, S., 'The law of the sea conference — recent developments, present status, and future implications', Committee for Coordination of Joint Prospecting for Mineral Resources in Asian Offshore Areas (CCOP), 12th session, 8-22 August 1975, Tokyo, Japan.

Petroleum Economist, Vol. XL, No. 8 (August 1973), p. 311.

————, Vol. XL, No. 10 (October 1973), p. 373.

————, Vol. XLI, No. 10 (October 1974), p. 391.

————, Vol. XLIII, No. 3 (March 1976), p. 114.

————, Vol. XLIII, No. 7 (July 1976), p. 279.

————, Vol. XLIII, No. 9 (September 1976), p. 362.

————, Vol. XLIII, No. 10 (October 1976), p. 400.

————, Vol. XLIV, No. 7 (July 1977), p. 283.

Petroleum News S.E.A. (Hong Kong), Vol. 4, No. 3 (June 1973), p. 20.

————, Vol. 4, No. 4 (July 1973), p. 31.

————, Vol. 5, No. 2 (May 1974), p. 5.

————, Vol. 5, No. 3 (June 1974), p. 9.

————, Vol. 6, No. 3 (June 1975), pp. 3, 20.

————, Vol. 7, No. 11 (February 1977), p. 7.

POLOMKA, P., *Asean and the Law of the Sea*, Occasional Paper No. 36, Singapore: Institute of Southeast Asian Studies, 1975.

SCHACHTER, O. and SERVER, D., *Marine Pollution Problems and Remedies*, New York: U.N. Institute for Training and Research, 1971.

SINGH, HARBAJAN, 'Law of the sea and Southeast Asian problems', *Southeast Asian Affairs 1976*, Singapore: Institute of Southeast Asian Studies, 1976.

South China Morning Post (Hong Kong), 'Islands of multiple claims', 16 June 1976, reproduced in *The Mirror* (Singapore), 19 July 1976.

Straits Times (Singapore), 'More support for 12-mile limit, seabed rights', 26 August 1974, p. 1.

————, 'Singapore will lose out in new sea law', 14 July 1975, p. 1.

————, 'Spratlys: two questions', 30 November 1975, p. 12.

————, 'Peking warns: hands off Spratly Isles', 16 June 1976, p. 2.

————, 'Spratly Isles: drilling goes on despite China warning', 18 June 1976, p. 2.

————, 'Philippines can defend Spratly interest', 19 June 1976, p. 3.

————, 'Romulo for China talks on Spratlys?' 21 June 1976, p. 4.

————, 'Oil wealth sparks a dispute over Spratlys', 6 July 1976, p. 8.

————, 'The fuse that could set off an international explosion', 19 July 1976, p. 16.

————, 'Manila ready to renew claims', 22 July 1976, p. 4.

————, 'Sea lanes for foreign warships plan by Jakarta', 30 July 1976, p. 2.

————, 'UN law of sea talks likely to back free transit', 2 August 1976, p. 7.

————, 'UN's longest-running conference meets again on the law of the sea', 2 August 1976, p. 12.

————, 'Burma: It's time to protect maritime interests', 11 October 1976, p. 3.

————, 'Sea law: bid by KL to amend text', 14 May 1977, p. 20.

————, 'UN's broad pact to carve up world's oceans', 13 July 1977, p. 2.

————, 'Sea law talks end without pact', 17 July 1977, p. 2.

Sunday Nation (*Singapore*), 'Viets make claim on offshore zones', 22 May 1977, p. 4.

SWEENEY, R.J., TOLLISON, R.D., and WILLETT, T.D., 'Market failure, the common pool problem, and ocean resource exploitation', *Journal of Law and Economics*, Vol. 17, No. 1 (April 1974), pp. 179-92.

SWING, J.T., 'Who will own the oceans?', *Foreign Affairs*, April 1976, pp. 528-46.

United Nations, Committee on the Peaceful Uses of the Sea-Bed and the Ocean Floor Beyond the Limits of National Jurisdiction, 'Economic significance in terms of seabed mineral resources of the various limits proposed for national jurisdiction', Document A/AC. 138/87, 4 June 1973.

————, Legislative Series, *National Legislation and Treaties Relating to the Territorial Sea, the Contiguous Zone, the Continental Shelf, the High Seas, and to Fishing and Conservation of the Living Resources of the Sea*, New York, 1970.

————, *Report of the Committee on the Peaceful Uses of the Sea-Bed and the Ocean Floor Beyond the Limits of National Jurisdiction*, General Assembly Official Records, 24th Session, 1969; 25th Session, 1970; 26th Session, 1971; 27th Session, 1972.

————, Third Conference on the Law of the Sea, *Official Records*, Vols. I-III, First and Second Sessions, New York: 1975.

————, Third Conference on the Law of the Sea, *Official Records*, Vol. VI, Fifth Session, New York: 1977.

————, Third Conference on the Law of the Sea, 'Revised single negotiating text', Parts I, II, and III, A/CONF. 62/WP.8/Rev. 1, 6 May 1976.

————, Third Conference on the Law of the Sea, 'Revised single negotiating text', Part IV, Document A/CONF.62/WP.9/Rev. 2, 23 November 1976.

————, Third Conference on the Law of the Sea, 'Informal

composite negotiating text', Document A/CONF.62/WP.10/ Corr. 2, 20 July 1977.

United States, Congress, Committee on Foreign Affairs (House) and Committee on Foreign Relations (Senate), *Legislation on Foreign Relations*, Washington, D.C.: U.S. Government Printing Office, 1974.

————, Congress, House of Representatives, Committee on Foreign Affairs, *Oil and Asian Rivals*, 93rd Congress, 1st and 2nd sessions, Washington, D.C.: U.S. Government Printing Office, 1974.

————, Department of State, Bureau of Intelligence Research, Office of the Geographer, *Limits in the Seas Series, No. 36: National Claims to Maritime Jurisdictions*, third revision, Washington, D.C., 23 December 1975.

————, Presidential Proclamation No. 2667 on the 'Policy of the United States with respect to the natural resources of the subsoil and seabed of the continental shelf' (the Truman Proclamation), 28 September 1945.

VARON, B., 'Slow sailing at law of the sea: the implications for the future', *Finance and Development*, Vol. 12, No. 1 (March 1975), pp. 38-41.

Working People's Daily (Rangoon), 'Territorial sea and maritime zones law promulgated', 10 April 1977, pp. 1, 4.

Philippine Claim to Sabah

ARIFF, M.O., *The Philippines' Claim to Sabah*, Singapore—Kuala Lumpur: Oxford University Press, 1970.

JAYAKUMAR, S., 'Philippine claim to Sabah and international law', *Malaya Law Review*, Vol. 10, No. 2 (1968), pp. 306-35.

LEIFER, M., *The Philippine Claims on Sabah*, Hull: University of Hull, Centre for South-East Asian Studies, 1968.

Malaysia, Department of Information, *Philippine Grab Bill*, Speech of Prime Minister before Parliament, 15 October 1968 (pamphlet).

ORTIZ, P. (S.J.), *Legal Aspects of the North Borneo Question*, Manila: Bureau of Printing, 1964.

Petroleum News S.E.A., September 1976 News Supplement, annotated centrefold map.

Philippine Government, *Philippine Claim to Sabah*, Vols. I and II, Manila: Bureau of Printing, 1963.

————, Philippine Policy Statement in the United Nations, delivered by Narciso Ramos, before General Assembly, 23rd session, New York, 15 October 1968 (pamphlet).

Straits Times (Singapore), 'Marcos: why coming ASEAN summit is critical', 29 January 1976, p. 12.

————, 'Philippines to drop claim to Sabah?', 16 July 1977, p. 3.

The Sunday Mail (Kuala Lumpur), 'The Sabah claim', 8 July 1973 and 15 July 1973.

Special Petroleum, Technical, Legal and Other Studies

CARLSON, S., *Indonesia's Oil*, Washington, D.C.: Georgetown University, Center for Strategic Studies, 1976.

CLEWELL, D.H., 'Advantages of cooperative research for the oil industry—what could API's role be?', American Petroleum Institute Division of Production Meeting, Houston, Texas, March 1972, manuscript.

DURKEE, F.F. and HATLEY, A.G., 'The Philippines: is a second exploration cycle warranted?', *The Oil and Gas Journal*, 18 January 1971, pp. 86-9.

GAFFNEY, P.D., MOYES, C.P., and ALING, B., 'Economic appraisal of the potential petroleum resources of the Asian Pacific region', Offshore Southeast Asia Conference, SPE session, Singapore, 19 February 1976.

GROSSLING, B.F., *Latin America's Petroleum Prospects in the Energy Crisis*, U.S. Geological Survey Bulletin 1411, Washington, D.C.: U.S. Government Printing Office, 1975.

HAMMOND, A.I., 'Bright spot: better seismological indicators of gas and oil', *Science*, Vol. 185 (9 August 1974), pp. 515-17.

HATLEY, A.G., 'Offshore petroleum exploration in East Asia—an overview', Proceedings of Offshore Southeast Asia Conference,

SEAPEX Program, Paper 1, Singapore, 1976.

HOWELL, L. and MORROW, M., *Asia, Oil Politics, and the Energy Crisis*, IDOC/International Documentation Nos. 60-61, New York: IDOC/North America, 1974.

MICHIE, M.S., 'The search for oil in Brunei', *Petroleum di Brunei* (Brunei Shell Petroleum Company, April 1975).

Mitre Corporation and Malaysian Government, *Fifth International Symposium on Energy, Resources and the Environment*, Kuala Lumpur, 1975.

National Petroleum Council, *Ocean Petroleum Resources*, Washington, D.C.: 1975.

NG, S.M., *The Oil System in Southeast Asia*, Field Report Series, No. 8, Singapore: Institute of Southeast Asian Studies, 1974.

Oriental Petroleum and Minerals Corporation, *Annual Report, 1975*, Manila.

RAJARETNAM, M., *Politics of Oil in the Philippines*, Field Report Series, No. 6, Singapore: Institute of Southeast Asian Studies, 1973.

SIDDAYAO, C.M., 'Looking for oil in the Philippines', *Esso Eastern Review* (New York), June 1966, pp. 3-5.

TENGKU TAN SRI RAZALEIGH HAMZAH, 'Oil industry and its impact on Malaysian national development', *Indonesia Oil and Gas* (Singapore), Vol. 2, No. 3 (October 1976), pp. 3-6.

United Nations, Economic Commission for Asia and the Far East, *Case Histories of Oil and Gas Fields in Asia and the Far East*, New York, 1963, 1967, and 1971, Mineral Resources Development Series, Nos. 20, 29 and 37.

————, *Mineral Resources of the Lower Mekong Basin and Adjacent Areas of Khmer Republic, Laos, Thailand and Republic of Viet-Nam*, Mineral Resources Development Series No. 39.

————, *Proceedings of the Symposium on the Development of Petroleum Resources of Asia and the Far East*, Mineral Resources Development Series No. 10, New York: 1959.

————, *Proceedings of the Second Symposium on the Development of Petroleum Resources of Asia and the Far East*, Mineral Resources Development Series No. 18, Vols. I, II, and III, New York: 1963.

————, *Proceedings of the Third Symposium on the Development of Petroleum Resources of Asia and the Far East*, Mineral Resources Development Series No. 26, Vols. I, II and III, New York: 1967.

————, *Proceedings of the Fourth Symposium on the Development of Petroleum Resources of Asia and the Far East*, Mineral Resources Development Series No. 43, New York: 1973.

————, Economic and Social Commission for Asia and the Pacific, *Proceedings of the Twelfth Session of the Sub-Committee on Energy Resources and Electric Power*, New York: 1974.

————, Economic Commission for Asia and the Far East, Committee for Coordination of Joint Prospecting for Mineral Resources in Asian Offshore Areas (CCOP), *Reports* of the Fifth, Sixth, and Ninth Sessions, Bangkok: 1968, 1969 and 1972.

————, Economic and Social Commission for Asia and the Pacific, CCOP, *Reports* of the Tenth, Eleventh, and Twelfth Sessions, Bangkok: 1973, 1974, 1975.

————, *The Offshore Hydrocarbon Potential of East Asia: A Review of Investigations, 1966-1973*, Bangkok: 1974.

————, *Technical Bulletin*, Vol. 2, issued May 1969, Vol. 3, issued May 1970, Bangkok.

United States, Council on Environmental Quality, *OCS Oil and Gas—An Environmental Assessment*, Vols. 1-5, Washington, D.C.: Government Printing Office, April 1974.

————, Department of Interior, Bureau of Mines, *Offshore Petroleum Studies*, Bureau of Mines Information Circular 8557, Washington, D.C.: U.S. Government Printing Office, 1972.

————, Federal Energy Administration, *The Relationship of Oil Companies and Foreign Governments*, Washington, D.C.: U.S. Government Printing Office, 1975.

————, Senate, Committee on Commerce, *Outer Continental Shelf Oil and Gas Development and the Coastal Zone*, Washington, D.C.: U.S. Government Printing Office, November 1974.

WARTELLE, T.R. and LEE, G.C., 'Fixed platform design for South East Asia', *Proceedings*, Offshore Southeast Asia Conference, SPE session, Singapore, 20 February 1976.

Other Data Sources

Periodicals

Asian Wall Street Journal (Hong Kong), 'ASEAN agrees to share fuel in emergency', 18 October 1976, pp. 1, 5.

————, 'Big oil vs. new Malaysian nationalism', 3 November 1976, p. 4.

————, 'Brunei: a British anachronism', 9 November 1976, p. 4.

————, 'Singapore's economy is entering a phase of reduced growth', 17 February 1977, p. 1, 7.

————, 'Exxon unit drills at record depths off Thailand', 28 January 1977, p. 3.

————, 'Thailand says gas-price accord near, but a U.S. producer is more cautious', 1 September 1977, p. 32.

Bangkok Post, 'Burma oil output increases by 2.6%', 16 March 1976, p. 10.

Business Week (U.S.), 'The worldwide search for oil', 3 February 1975, pp. 38-44.

Economist Intelligence Unit (London), *Quarterly Economic Review—Indonesia*, Annual Supplement, 1976.

————, *Quarterly Economic Review—Oil in the Far East and Australasia*, various issues.

Far Eastern Economic Review (Hong Kong), 'Tapping Burma's on-shore oil', 25 June 1976, p. 84.

KRAAR, L., 'Tomorrow casts a shadow on an Asian Eden', *Fortune*, August 1976, pp. 192-4.

LIM, J. J., 'Brunei: prospects for a "Protectorate" ', *Southeast Asian Affairs 1976*, Singapore: Institute of Southeast Asian Studies, 1976, pp. 149-64.

Oil and Gas Journal, Mid-year and year-end issues on production, reserves and resources.

Petroleum Economist (London), Vol. XLIII, No. 1 (January 1976), and No. 2 (February 1976).

Petroleum News S.E.A., 'Brunei: Southeast Asia's pocket-size producer', Vol. 4, No. 4 (July 1973), pp. 38-41.

————, 'Sabah on ... but slowly', Vol. 7, No. 5, August 1976, p. 16.

Straits Times (Singapore), 'Straits: joint policy on pollution', 25 February 1977, p. 11.
World Oil (U.S.), Annual international outlook issue.

Statistical Reports and Books

ALBER, J.P. *et al.*, *Summary Petroleum and Selected Mineral Statistics for 120 Countries Including Offshore Areas*, Geological Survey Professional Paper 817, Washington, D.C.: U.S. Government Printing Office, 1973.
American Petroleum Institute *et al.*, *Joint Association Survey of the U.S. Oil and Gas Producing Industry*, Annual.
————, *Quarterly Review of Drilling Statistics for the United States*, Washington, D.C.
British Petroleum Company, *BP Statistical Review of the World Oil Industry*, Annual.
Brunei, Economic Planning Unit, *Statistics of External Trade*, 1973.
Burma, Socialist Republic of the Union of, Ministry of Planning and Finance, *Report to the Pyithu Hluttaw on the Financial, Economic and Social Conditions of the Socialist Republic of the Union of Burma for 1975-76*, Rangoon, 1975.
Chase Manhattan Bank, *Capital Investments of the World Petroleum Industry*, Annual.
FREZON, SHERWOOD, *Summary of 1972 Oil and Gas Statistics for Onshore and Offshore Areas of 151 Countries*, Geological Survey Professional Paper 885, Washington, D.C.: U.S. Government Printing Office, 1974.
Indonesia, Bura Pusat Statistik, *Statistik Indonesia*, 1972/73 and 1974/75 issues, Jakarta.
International Monetary Fund, *International Financial Statistics*, May 1974 and July 1976 issues, Washington, D.C.
International Petroleum Encyclopedia, Tulsa, Oklahoma: The Petroleum Publishing Company, Annual.
Malaysia, Bank Negara Malaysia, *Quarterly Economic Bulletin*, March-June 1975.
Oil and Gas International Yearbook, London: Financial Times, Annual.

Thailand, Department of Mineral Resources, Ministry of Industry, *Petroleum Activities in Thailand,* Bangkok, March 1976.

———, ESCAP Thailand Delegation, Committee on Natural Resources, *Progress Report on Energy Development in Thailand,* presented to Economic and Social Commission for Asia and the Pacific, Committee on National Resources, October 1975.

United Nations, Economic Commission for Asia and the Far East, *Statistical Yearbook for Asia and the Far East,* Annual.

———, *World Energy Supplies, 1950-1974,* Statistical Papers, Series J, No. 19, New York, 1976.

United States, Department of Interior, Bureau of Mines, *Mineral Industry Surveys:* 'World Crude Oil Production', Annual, 1975, issued 9 June 1976.

———, State Department, Embassy, Jakarta, *Indonesian Petroleum Report, 1973/74* (issued July 1974), 1975 (issued June 1975) and 1976 (issued June 1976).

———, Office of the President, *International Economic Report of the President,* Washington, D.C.: Government Printing Office, 1976.

World Bank, *Atlas,* Washington, D.C.: Annual.

Index

ABYSSAL plains, 35n
Aegean Sea, Greek/Turkish dispute, 106, 168
Agreements, Chile-Ecuador-Peru, 70n (see also Indonesia, Malaysia, Singapore, and Thailand)
Andaman Sea: 63; agreements on shelf, 65; conflict potential, 110; exploration, 34; petroleum prospects, 29
Archipelagic principle: 56−8; Indonesian position, 43, 47−8, 54, 56; Philippine position, 43, 49−50, 54, 56, 70n
Archipelago: Composite Text, 71n, 174; USGS definition, 171
Ariff, M.O., 96−7
ASCOPE, 107, 168

BAGASSE as fuel, 13, 18n
Benefit/cost analysis and conflicts, 149−53
Borneo: continental shelf, 25; offshore exploration, 29; petroleum production, 29 (see also Brunei, Sabah, Sarawak)
'Bright spot' technique, 124, 157n
Brunei: continental shelf, 43, 46; energy consumption, 5, 9; offshore production, 32, 34; petroleum consumption, 9, 13, 15, 17; petroleum discoveries, 22, 28; petroleum exports: earnings, 16, 133; petroleum production, 32; petroleum resources, 26, 29; petroleum revenues, 16
Burma: continental shelf issues, 46, 63, 66; drilling, 22, 30, 37n, 126; economic zone, 46,66; energy con-

sumption, 5, 10, marine scientific research, 46, 106−7; petroleum consumption, 10, 13, 15, 17; petroleum production, 14, 32; petroleum resources, 26, 32; territorial sea issues, 46, 54

CAMBODIA: boundary disputes with Thailand and Vietnam, 76−83; continental shelf issues, 46, 58, 64, 66; Continental Shelf Convention, 77−80; energy consumption, 5, 10; energy/GNP ratio, 5, 18n; offshore exploration, 77−9; petroleum consumption, 10, 13, 15, 17; petroleum resources, 26; territorial sea issues, 44, 56
Capital investments: development and production, 123, 125−31, 138−40; Indonesia, 139−40, 160n; producing capacity, 131 (see also Costs)
CCOP: 105, 168; and marine pollution control, 112; and regional cooperation in research, 107; seismic survey of South-East Asian waters, 28, 29, 106
Charcoal as fuel, 13, 18n
China territorial claims: Spratlys, 84−7, 89−92; Paracels, 93−4
Cloma, Tomas, and Spratlys, 88−9
Coal consumption, South-East Asia, 9−12, 13
'Common heritage of mankind' concept, 44, 46−53
'Common pool' problem, 107−9; and conflicts settlement, 156; and the social discount rate, 110
Comparative advantage theory and

SOUTH-EAST ASIA

0 500 1000 KM

130°   140°   150°

20°

P A C I F I C

PHILIPPINES

O C E A N

10°

0°

A

E   S   I

130°   140°   150°